PREFACE

Introduction

Internationally, code officials recognize the need for a modern, up-to-date fuel gas code addressing the design and installation of fuel gas systems and gas-fired appliances through requirements emphasizing performance. The *International Fuel Gas Code*®, in this 2006 edition, is designed to meet these needs through model code regulations that safeguard the public health and safety in all communities, large and small.

This comprehensive fuel gas code establishes minimum regulations for fuel gas systems and gas-fired appliances using prescriptive and performance-related provisions. It is founded on broad-based principles that make possible the use of new materials and new fuel gas system and appliance designs. This 2006 edition is fully compatible with all the *International Codes*® (I-Codes®) published by the International Code Council (ICC)®, including the *International Building Code*®, ICC *Electrical Code*®, *International Energy Conservation Code*®, *International Existing Building Code*®, *International Fire Code*®, *International Mechanical Code*®, ICC *Performance Code*®, *International Plumbing Code*®, *International Private Sewage Disposal Code*®, *International Property Maintenance Code*®, *International Residential Code*®, *International Wildland-Urban Interface Code*™ and *International Zoning Code*®.

The *International Fuel Gas Code* provisions provide many benefits, among which is the model code development process that offers an international forum for fuel gas technology professionals to discuss performance and prescriptive code requirements. This forum provides an excellent arena to debate proposed revisions. This model code also encourages international consistency in the application of provisions.

Development

The first edition of the *International Fuel Gas Code* (1997) was the culmination of an effort initiated in 1996 by a development committee appointed by ICC and consisting of representatives of the three statutory members of the International Code Council at that time, including: Building Officials and Code Administrators International, Inc. (BOCA), International Conference of Building Officials (ICBO) and Southern Building Code Congress International (SBCCI) and the gas industry. The intent was to draft a comprehensive set of regulations for fuel gas systems and gas-fired appliances consistent with and inclusive of the scope of the existing mechanical, plumbing and gas codes. Technical content of the latest model codes promulgated by BOCA, ICBO, SBCCI and ICC and the *National Fuel Gas Code* (ANSI Z223.1) was utilized as the basis for the development. This 2006 edition presents the code as originally issued, with changes reflected in subsequent editions through 2003, and with code changes approved through the ICC Code Development Process through 2005 and standard revisions correlated with ANSI Z223.1-2006. A new edition such as this is promulgated every three years.

This code is founded on principles intended to establish provisions consistent with the scope of a fuel gas code that adequately protects public health, safety and welfare; provisions that do not unnecessarily increase construction costs; provisions that do not restrict the use of new materials, products or methods of construction; and provisions that do not give preferential treatment to particular types or classes of materials, products or methods of construction.

Format

The *International Fuel Gas Code* is segregated by section numbers into two categories — "code" and "standard" — all coordinated and incorporated into a single document. The sections that are "code" are designated by the acronym "IFGC" next to the main section number (e.g., Section 101). The sections that are "standard" are designated by the acronym "IFGS" next to the main section number (e.g., Section 304).

Adoption

The *International Fuel Gas Code* is available for adoption and use by jurisdictions internationally. Its use within a governmental jurisdiction is intended to be accomplished through adoption by reference in accordance with proceedings establishing the jurisdiction's laws. At the time of adoption, jurisdictions should insert the appropriate information in provisions requiring specific local information, such as the name of the adopting jurisdiction. These locations are shown in bracketed words in small capital letters in the code and in the sample ordinance. The sample adoption ordinance on page v addresses several key elements of a code adoption ordinance, including the information required for insertion into the code text.

Maintenance

The *International Fuel Gas Code* is kept up to date through the review of proposed changes submitted by code enforcing officials, industry representatives, design professionals and other interested parties. Proposed changes are carefully considered through an open code development process in which all interested and affected parties may participate. The code development process of the *International Fuel Gas Code* is slightly different than the process for the other *International Codes.*

Proposed changes to text designated "IFGC" are subject to the ICC Code Development Process. For more information regarding the code development process, contact the Code and Standard Development Department of the International Code Council.

Proposed changes to text designated as "IFGS" are subject to the standards development process which maintains the *National Fuel Gas Code* (ANSI Z223.1). For more information regarding the standard development process, contact the American Gas Association (AGA) at 400 N. Capitol Street, N.W., Washington, DC 20001.

While the development procedure of the *International Fuel Gas Code* assures the highest degree of care, the ICC, its members, the AGA and those participating in the development of this code do not accept any liability resulting from compliance or noncompliance with the provisions because the ICC, its founding members and the AGA do not have the power or authority to police or enforce compliance with the contents of this code. Only the governmental body that enacts the code into law has such authority.

Letter Designations in Front of Section Numbers

In each code development cycle, proposed changes to the code are considered at the Code Development Hearings by the ICC Fuel Gas Code Development Committee, whose action constitutes a recommendation to the voting membership for final action on the proposed change. Proposed changes to a code section that has a number beginning with a letter in brackets are considered by a different code development committee. For example, proposed changes to code sections that have [B] in front of them (e.g., [B]302.1) are considered by the International Building Code Development Committee at the code development hearings.

The content of sections in this code that begin with a letter designation are maintained by another code development committee in accordance with the following:

[B] = International Building Code Development Committee;

[M] = International Mechanical Code Development Committee; and

[F] = International Fire Code Development Committee.

Marginal Markings

Solid vertical lines in the margins within the body of the code indicate a technical change from the requirements of the 2003 edition. Deletion indicators in the form of an arrow (➡) are provided in the margin where an entire section, paragraph, exception or table has been deleted or item in a list of items or in a table has been deleted.

A Member of the International Code Family®

INTERNATIONAL

FUEL

GAS

CODE®

2006

2006 International Fuel Gas Code®

First Printing: January 2006
Second Printing: March 2006
Third Printing: February 2007
Fourth Printing: June 2007

ISBN-13: 978-1-58001-269-0 (soft)
ISBN-10: 1-58001-269-8 (soft)
ISBN-13: 978-1-58001-268-3 (loose leaf)
ISBN-10: 1-58001-268-x (loose leaf)
ISBN-13: 978-1-58001-305-5 (e-document)
ISBN-10: 1-58001-305-8 (e-document)

Material designated IFGS
by
AMERICAN GAS ASSOCIATION
400 N. Capitol Street, N.W. • Washington, DC 20001
(202) 824-7000
Copyright © American Gas Association, 2006. All rights reserved.

PRINTED IN THE U.S.A.

ORDINANCE

The *International Codes* are designed and promulgated to be adopted by reference by ordinance. Jurisdictions wishing to adopt the 2006 *International Fuel Gas Code* as an enforceable regulation governing fuel gas systems and gas-fired appliances should ensure that certain factual information is included in the adopting ordinance at the time adoption is being considered by the appropriate governmental body. The following sample adoption ordinance addresses several key elements of a code adoption ordinance, including the information required for insertion into the code text.

SAMPLE ORDINANCE FOR ADOPTION OF THE *INTERNATIONAL FUEL GAS CODE* ORDINANCE NO._____

An ordinance of the **[JURISDICTION]** adopting the 2006 edition of the *International Fuel Gas Code*, regulating and governing fuel gas systems and gas-fired appliances in the **[JURISDICTION]**; providing for the issuance of permits and collection of fees therefor; repealing Ordinance No. _____ of the **[JURISDICTION]** and all other ordinances and parts of the ordinances in conflict therewith.

The **[GOVERNING BODY]** of the **[JURISDICTION]** does ordain as follows:

Section 1. That a certain document, three (3) copies of which are on file in the office of the **[TITLE OF JURISDICTION'S KEEPER OF RECORDS]** of **[NAME OF JURISDICTION]** , being marked and designated as the *International Fuel Gas Code*, 2006 edition, including Appendix Chapters **[FILL IN THE APPENDIX CHAPTERS BEING ADOPTED]** (see *International Fuel Gas Code* Section 101.3, 2006 edition), as published by the International Code Council, be and is hereby adopted as the Fuel Gas Code of the **[JURISDICTION]**, in the State of **[STATE NAME]** for regulating and governing fuel gas systems and gas-fired appliances as herein provided; providing for the issuance of permits and collection of fees therefor; and each and all of the regulations, provisions, penalties, conditions and terms of said Fuel Gas Code on file in the office of the **[JURISDICTION]** are hereby referred to, adopted, and made a part hereof, as if fully set out in this ordinance, with the additions, insertions, deletions and changes, if any, prescribed in Section 2 of this ordinance.

Section 2. The following sections are hereby revised:

Section 101.1. Insert: **[NAME OF JURISDICTION]**

Section 106.5.2. Insert: **[APPROPRIATE SCHEDULE]**

Section 106.5.3. Insert: **[PERCENTAGES IN TWO LOCATIONS]**

Section 108.4. Insert: **[SPECIFY OFFENSE] [AMOUNT] [NUMBER OF DAYS]**

Section 108.5. Insert: **[AMOUNTS IN TWO LOCATIONS]**

Section 3. That Ordinance No. _____ of **[JURISDICTION]** entitled **[FILL IN HERE THE COMPLETE TITLE OF THE ORDINANCE OR ORDINANCES IN EFFECT AT THE PRESENT TIME SO THAT THEY WILL BE REPEALED BY DEFINITE MENTION]** and all other ordinances or parts of ordinances in conflict herewith are hereby repealed.

Section 4. That if any section, subsection, sentence, clause or phrase of this ordinance is, for any reason, held to be unconstitutional, such decision shall not affect the validity of the remaining portions of this ordinance. The **[GOVERNING BODY]** hereby declares that it would have passed this ordinance, and each section, subsection, clause or phrase thereof, irrespective of the fact that any one or more sections, subsections, sentences, clauses and phrases be declared unconstitutional.

Section 5. That nothing in this ordinance or in the Fuel Gas Code hereby adopted shall be construed to affect any suit or proceeding impending in any court, or any rights acquired, or liability incurred, or any cause or causes of action acquired or existing, under any act or ordinance hereby repealed as cited in Section 3 of this ordinance; nor shall any just or legal right or remedy of any character be lost, impaired or affected by this ordinance.

Section 6. That the **[JURISDICTION'S KEEPER OF RECORDS]** is hereby ordered and directed to cause this ordinance to be published. (An additional provision may be required to direct the number of times the ordinance is to be published and to specify that it is to be in a newspaper in general circulation. Posting may also be required.)

Section 7. That this ordinance and the rules, regulations, provisions, requirements, orders and matters established and adopted hereby shall take effect and be in full force and effect **[TIME PERIOD]** from and after the date of its final passage and adoption.

TABLE OF CONTENTS

CHAPTER 1

ADMINISTRATION

SECTION 101 (IFGC)
GENERAL

101.1 Title. These regulations shall be known as the *Fuel Gas Code* of [NAME OF JURISDICTION], hereinafter referred to as "this code."

101.2 Scope. This code shall apply to the installation of fuel gas piping systems, fuel gas utilization equipment, gaseous hydrogen systems and related accessories in accordance with Sections 101.2.1 through 101.2.5.

Exception: Detached one- and two-family dwellings and multiple single-family dwellings (townhouses) not more than three stories high with separate means of egress and their accessory structures shall comply with the *International Residential Code.*

101.2.1 Gaseous hydrogen systems. Gaseous hydrogen systems shall be regulated by Chapter 7.

101.2.2 Piping systems. These regulations cover piping systems for natural gas with an operating pressure of 125 pounds per square inch gauge (psig) (862 kPa gauge) or less, and for LP-gas with an operating pressure of 20 psig (140 kPa gauge) or less, except as provided in Section 402.6.1. Coverage shall extend from the point of delivery to the outlet of the equipment shutoff valves. Piping systems requirements shall include design, materials, components, fabrication, assembly, installation, testing, inspection, operation and maintenance.

101.2.3 Gas utilization equipment. Requirements for gas utilization equipment and related accessories shall include installation, combustion and ventilation air and venting and connections to piping systems.

101.2.4 Systems and equipment outside the scope. This code shall not apply to the following:

1. Portable LP-gas equipment of all types that is not connected to a fixed fuel piping system.

2. Installation of farm equipment such as brooders, dehydrators, dryers and irrigation equipment.

3. Raw material (feedstock) applications except for piping to special atmosphere generators.

4. Oxygen-fuel gas cutting and welding systems.

5. Industrial gas applications using gases such as acetylene and acetylenic compounds, hydrogen, ammonia, carbon monoxide, oxygen and nitrogen.

6. Petroleum refineries, pipeline compressor or pumping stations, loading terminals, compounding plants, refinery tank farms and natural gas processing plants.

7. Integrated chemical plants or portions of such plants where flammable or combustible liquids or gases are produced by, or used in, chemical reactions.

8. LP-gas installations at utility gas plants.

9. Liquefied natural gas (LNG) installations.

10. Fuel gas piping in power and atomic energy plants.

11. Proprietary items of equipment, apparatus or instruments such as gas-generating sets, compressors and calorimeters.

12. LP-gas equipment for vaporization, gas mixing and gas manufacturing.

13. Temporary LP-gas piping for buildings under construction or renovation that is not to become part of the permanent piping system.

14. Installation of LP-gas systems for railroad switch heating.

15. Installation of hydrogen gas, LP-gas and compressed natural gas (CNG) systems on vehicles.

16. Except as provided in Section 401.1.1, gas piping, meters, gas pressure regulators and other appurtenances used by the serving gas supplier in the distribution of gas, other than undiluted LP-gas.

17. Building design and construction, except as specified herein.

18. Piping systems for mixtures of gas and air within the flammable range with an operating pressure greater than 10 psig (69 kPa gauge).

19. Portable fuel cell appliances that are neither connected to a fixed piping system nor interconnected to a power grid.

101.2.5 Other fuels. The requirements for the design, installation, maintenance, alteration and inspection of mechanical systems operating with fuels other than fuel gas shall be regulated by the *International Mechanical Code.*

101.3 Appendices. Provisions in the appendices shall not apply unless specifically adopted.

101.4 Intent. The purpose of this code is to provide minimum standards to safeguard life or limb, health, property and public welfare by regulating and controlling the design, construction, installation, quality of materials, location, operation and maintenance or use of fuel gas systems.

101.5 Severability. If a section, subsection, sentence, clause or phrase of this code is, for any reason, held to be unconstitutional, such decision shall not affect the validity of the remaining portions of this code.

SECTION 102 (IFGC)
APPLICABILITY

102.1 General. The provisions of this code shall apply to all matters affecting or relating to structures and premises, as set forth in Section 101. Where, in a specific case, different sections of this code specify different materials, methods of construction or other requirements, the most restrictive shall govern.

102.2 Existing installations. Except as otherwise provided for in this chapter, a provision in this code shall not require the removal, alteration or abandonment of, nor prevent the continued utilization and maintenance of, existing installations lawfully in existence at the time of the adoption of this code.

> **[EB] 102.2.1 Existing buildings.** Additions, alterations, renovations or repairs related to building or structural issues shall be regulated by the *International Building Code.*

102.3 Maintenance. Installations, both existing and new, and parts thereof shall be maintained in proper operating condition in accordance with the original design and in a safe condition. Devices or safeguards which are required by this code shall be maintained in compliance with the code edition under which they were installed. The owner or the owner's designated agent shall be responsible for maintenance of installations. To determine compliance with this provision, the code official shall have the authority to require an installation to be reinspected.

102.4 Additions, alterations or repairs. Additions, alterations, renovations or repairs to installations shall conform to that required for new installations without requiring the existing installation to comply with all of the requirements of this code. Additions, alterations or repairs shall not cause an existing installation to become unsafe, hazardous or overloaded.

Minor additions, alterations, renovations and repairs to existing installations shall meet the provisions for new construction, unless such work is done in the same manner and arrangement as was in the existing system, is not hazardous and is approved.

102.5 Change in occupancy. It shall be unlawful to make a change in the occupancy of a structure which will subject the structure to the special provisions of this code applicable to the new occupancy without approval. The code official shall certify that such structure meets the intent of the provisions of law governing building construction for the proposed new occupancy and that such change of occupancy does not result in any hazard to the public health, safety or welfare.

102.6 Historic buildings. The provisions of this code relating to the construction, alteration, repair, enlargement, restoration, relocation or moving of buildings or structures shall not be mandatory for existing buildings or structures identified and classified by the state or local jurisdiction as historic buildings when such buildings or structures are judged by the code official to be safe and in the public interest of health, safety and welfare regarding any proposed construction, alteration, repair, enlargement, restoration, relocation or moving of buildings.

102.7 Moved buildings. Except as determined by Section 102.2, installations that are a part of buildings or structures moved into or within the jurisdiction shall comply with the provisions of this code for new installations.

102.8 Referenced codes and standards. The codes and standards referenced in this code shall be those that are listed in Chapter 8 and such codes and standards shall be considered part of the requirements of this code to the prescribed extent of each such reference. Where differences occur between provisions of this code and the referenced standards, the provisions of this code shall apply.

> **Exception:** Where enforcement of a code provision would violate the conditions of the listing of the equipment or appliance, the conditions of the listing and the manufacturer's installation instructions shall apply.

102.9 Requirements not covered by code. Requirements necessary for the strength, stability or proper operation of an existing or proposed installation, or for the public safety, health and general welfare, not specifically covered by this code, shall be determined by the code official.

SECTION 103 (IFGC)
DEPARTMENT OF INSPECTION

103.1 General. The Department of Inspection is hereby created and the executive official in charge thereof shall be known as the code official.

103.2 Appointment. The code official shall be appointed by the chief appointing authority of the jurisdiction; and the code official shall not be removed from office except for cause and after full opportunity to be heard on specific and relevant charges by and before the appointing authority.

103.3 Deputies. In accordance with the prescribed procedures of this jurisdiction and with the concurrence of the appointing authority, the code official shall have the authority to appoint a deputy code official, other related technical officers, inspectors and other employees.

103.4 Liability. The code official, officer or employee charged with the enforcement of this code, while acting for the jurisdiction, shall not thereby be rendered liable personally, and is hereby relieved from all personal liability for any damage accruing to persons or property as a result of an act required or permitted in the discharge of official duties.

Any suit instituted against any officer or employee because of an act performed by that officer or employee in the lawful discharge of duties and under the provisions of this code shall be defended by the legal representative of the jurisdiction until the final termination of the proceedings. The code official or any subordinate shall not be liable for costs in an action, suit or proceeding that is instituted in pursuance of the provisions of this code; and any officer of the Department of Inspection, acting in good faith and without malice, shall be free from liability for acts performed under any of its provisions or by reason of any act or omission in the performance of official duties in connection therewith.

SECTION 104 (IFGC)
DUTIES AND POWERS OF THE CODE OFFICIAL

104.1 General. The code official shall enforce the provisions of this code and shall act on any question relative to the installation, alteration, repair, maintenance or operation of systems, except as otherwise specifically provided for by statutory requirements or as provided for in Sections 104.2 through 104.8.

104.2 Rule-making authority. The code official shall have authority as necessary in the interest of public health, safety and general welfare to adopt and promulgate rules and regulations; interpret and implement the provisions of this code; secure the intent thereof and designate requirements applicable because of local climatic or other conditions. Such rules shall not have the effect of waiving structural or fire performance requirements specifically provided for in this code, or of violating accepted engineering methods involving public safety.

104.3 Applications and permits. The code official shall receive applications and issue permits for installations and alterations under the scope of this code, inspect the premises for which such permits have been issued and enforce compliance with the provisions of this code.

104.4 Inspections. The code official shall make all of the required inspections, or shall accept reports of inspection by approved agencies or individuals. All reports of such inspections shall be in writing and shall be certified by a responsible officer of such approved agency or by the responsible individual. The code official is authorized to engage such expert opinion as deemed necessary to report upon unusual technical issues that arise, subject to the approval of the appointing authority.

104.5 Right of entry. Whenever it is necessary to make an inspection to enforce the provisions of this code, or whenever the code official has reasonable cause to believe that there exists in a building or upon any premises any conditions or violations of this code that make the building or premises unsafe, dangerous or hazardous, the code official shall have the authority to enter the building or premises at all reasonable times to inspect or to perform the duties imposed upon the code official by this code. If such building or premises is occupied, the code official shall present credentials to the occupant and request entry. If such building or premises is unoccupied, the code official shall first make a reasonable effort to locate the owner or other person having charge or control of the building or premises and request entry. If entry is refused, the code official has recourse to every remedy provided by law to secure entry.

When the code official has first obtained a proper inspection warrant or other remedy provided by law to secure entry, an owner or occupant or person having charge, care or control of the building or premises shall not fail or neglect, after proper request is made as herein provided, to promptly permit entry therein by the code official for the purpose of inspection and examination pursuant to this code.

104.6 Identification. The code official shall carry proper identification when inspecting structures or premises in the performance of duties under this code.

104.7 Notices and orders. The code official shall issue all necessary notices or orders to ensure compliance with this code.

104.8 Department records. The code official shall keep official records of applications received, permits and certificates issued, fees collected, reports of inspections, and notices and orders issued. Such records shall be retained in the official records as long as the building or structure to which such records relate remains in existence, unless otherwise provided for by other regulations.

SECTION 105 (IFGC)
APPROVAL

105.1 Modifications. Whenever there are practical difficulties involved in carrying out the provisions of this code, the code official shall have the authority to grant modifications for individual cases, provided the code official shall first find that special individual reason makes the strict letter of this code impractical and that such modification is in compliance with the intent and purpose of this code and does not lessen health, life and fire safety requirements. The details of action granting modifications shall be recorded and entered in the files of the Department of Inspection.

105.2 Alternative materials, methods and equipment. The provisions of this code are not intended to prevent the installation of any material or to prohibit any method of construction not specifically prescribed by this code, provided that any such alternative has been approved. An alternative material or method of construction shall be approved where the code official finds that the proposed design is satisfactory and complies with the intent of the provisions of this code, and that the material, method or work offered is, for the purpose intended, at least the equivalent of that prescribed in this code in quality, strength, effectiveness, fire resistance, durability and safety.

105.3 Required testing. Whenever there is insufficient evidence of compliance with the provisions of this code, evidence that a material or method does not conform to the requirements of this code, or in order to substantiate claims for alternative materials or methods, the code official shall have the authority to require tests as evidence of compliance to be made at no expense to the jurisdiction.

105.3.1 Test methods. Test methods shall be as specified in this code or by other recognized test standards. In the absence of recognized and accepted test methods, the code official shall approve the testing procedures.

105.3.2 Testing agency. All tests shall be performed by an approved agency.

105.3.3 Test reports. Reports of tests shall be retained by the code official for the period required for retention of public records.

105.4 Material and equipment reuse. Materials, equipment and devices shall not be reused unless such elements have been reconditioned, tested and placed in good and proper working condition, and approved.

SECTION 106 (IFGC)
PERMITS

106.1 When required. An owner, authorized agent or contractor who desires to erect, install, enlarge, alter, repair, remove, convert or replace an installation regulated by this code, or to cause such work to be done, shall first make application to the code official and obtain the required permit for the work.

Exception: Where equipment replacements and repairs are required to be performed in an emergency situation, the permit application shall be submitted within the next working business day of the Department of Inspection.

106.2 Permits not required. Permits shall not be required for the following:

1. Any portable heating appliance.
2. Replacement of any minor component of equipment that does not alter approval of such equipment or make such equipment unsafe.

Exemption from the permit requirements of this code shall not be deemed to grant authorization for work to be done in violation of the provisions of this code or of other laws or ordinances of this jurisdiction.

106.3 Application for permit. Each application for a permit, with the required fee, shall be filed with the code official on a form furnished for that purpose and shall contain a general description of the proposed work and its location. The application shall be signed by the owner or an authorized agent. The permit application shall indicate the proposed occupancy of all parts of the building and of that portion of the site or lot, if any, not covered by the building or structure and shall contain such other information required by the code official.

106.3.1 Construction documents. Construction documents, engineering calculations, diagrams and other data shall be submitted in two or more sets with each application for a permit. The code official shall require construction documents, computations and specifications to be prepared and designed by a registered design professional when required by state law. Construction documents shall be drawn to scale and shall be of sufficient clarity to indicate the location, nature and extent of the work proposed and show in detail that the work conforms to the provisions of this code. Construction documents for buildings more than two stories in height shall indicate where penetrations will be made for installations and shall indicate the materials and methods for maintaining required structural safety, fire-resistance rating and fireblocking.

> **Exception:** The code official shall have the authority to waive the submission of construction documents, calculations or other data if the nature of the work applied for is such that reviewing of construction documents is not necessary to determine compliance with this code.

106.4 Permit issuance. The application, construction documents and other data filed by an applicant for a permit shall be reviewed by the code official. If the code official finds that the proposed work conforms to the requirements of this code and all laws and ordinances applicable thereto, and that the fees specified in Section 106.5 have been paid, a permit shall be issued to the applicant.

106.4.1 Approved construction documents. When the code official issues the permit where construction documents are required, the construction documents shall be endorsed in writing and stamped "APPROVED." Such approved construction documents shall not be changed, modified or altered without authorization from the code official. Work shall be done in accordance with the approved construction documents.

The code official shall have the authority to issue a permit for the construction of part of an installation before the construction documents for the entire installation have been submitted or approved, provided adequate information and detailed statements have been filed complying with all pertinent requirements of this code. The holder of such permit shall proceed at his or her own risk without assurance that the permit for the entire installation will be granted.

106.4.2 Validity. The issuance of a permit or approval of construction documents shall not be construed to be a permit for, or an approval of, any violation of any of the provisions of this code or of other ordinances of the jurisdiction. A permit presuming to give authority to violate or cancel the provisions of this code shall be invalid.

The issuance of a permit based upon construction documents and other data shall not prevent the code official from thereafter requiring the correction of errors in said construction documents and other data or from preventing building operations from being carried on thereunder when in violation of this code or of other ordinances of this jurisdiction.

106.4.3 Expiration. Every permit issued by the code official under the provisions of this code shall expire by limitation and become null and void if the work authorized by such permit is not commenced within 180 days from the date of such permit, or is suspended or abandoned at any time after the work is commenced for a period of 180 days. Before such work recommences, a new permit shall be first obtained and the fee, therefor, shall be one-half the amount required for a new permit for such work, provided no changes have been or will be made in the original construction documents for such work, and further that such suspension or abandonment has not exceeded one year.

106.4.4 Extensions. A permittee holding an unexpired permit shall have the right to apply for an extension of the time within which he or she will commence work under that permit when work is unable to be commenced within the time required by this section for good and satisfactory reasons. The code official shall extend the time for action by the permittee for a period not exceeding 180 days if there is reasonable cause. A permit shall not be extended more than once. The fee for an extension shall be one-half the amount required for a new permit for such work.

106.4.5 Suspension or revocation of permit. The code official shall revoke a permit or approval issued under the provisions of this code in case of any false statement or misrepresentation of fact in the application or on the construction documents upon which the permit or approval was based.

106.4.6 Retention of construction documents. One set of construction documents shall be retained by the code official until final approval of the work covered therein. One set of approved construction documents shall be returned to the applicant, and said set shall be kept on the site of the building or work at all times during which the work authorized thereby is in progress.

106.5 Fees. A permit shall not be issued until the fees prescribed in Section 106.5.2 have been paid, nor shall an amendment to a permit be released until the additional fee, if any, due to an increase of the installation, has been paid.

106.5.1 Work commencing before permit issuance. Any person who commences work on an installation before obtaining the necessary permits shall be subject to 100 percent of the usual permit fee in addition to the required permit fees.

106.5.2 Fee schedule. The fees for work shall be as indicated in the following schedule.

[JURISDICTION TO INSERT APPROPRIATE SCHEDULE]

106.5.3 Fee refunds. The code official shall authorize the refunding of fees as follows.

1. The full amount of any fee paid hereunder which was erroneously paid or collected.

2. Not more than [SPECIFY PERCENTAGE] percent of the permit fee paid when no work has been done under a permit issued in accordance with this code.

3. Not more than [SPECIFY PERCENTAGE] percent of the plan review fee paid when an application for a permit for which a plan review fee has been paid is withdrawn or canceled before any plan review effort has been expended.

The code official shall not authorize the refunding of any fee paid, except upon written application filed by the original permittee not later than 180 days after the date of fee payment.

SECTION 107 (IFGC)
INSPECTIONS AND TESTING

107.1 Required inspections and testing. The code official, upon notification from the permit holder or the permit holder's agent, shall make the following inspections and other such inspections as necessary, and shall either release that portion of the construction or notify the permit holder or the permit holder's agent of violations that are required to be corrected. The holder of the permit shall be responsible for scheduling such inspections.

1. Underground inspection shall be made after trenches or ditches are excavated and bedded, piping is installed and before backfill is put in place. When excavated soil contains rocks, broken concrete, frozen chunks and other rubble that would damage or break the piping or cause corrosive action, clean backfill shall be on the job site.

2. Rough-in inspection shall be made after the roof, framing, fireblocking and bracing are in place and components to be concealed are complete, and prior to the installation of wall or ceiling membranes.

3. Final inspection shall be made upon completion of the installation.

The requirements of this section shall not be considered to prohibit the operation of any heating equipment installed to replace existing heating equipment serving an occupied portion of a structure in the event a request for inspection of such heating equipment has been filed with the department not more than 48 hours after replacement work is completed, and before any portion of such equipment is concealed by any permanent portion of the structure.

107.1.1 Approved inspection agencies. The code official shall accept reports of approved agencies, provided that such agencies satisfy the requirements as to qualifications and reliability.

107.1.2 Evaluation and follow-up inspection services. Prior to the approval of a prefabricated construction assembly having concealed work and the issuance of a permit, the code official shall require the submittal of an evaluation report on each prefabricated construction assembly, indicating the complete details of the installation, including a description of the system and its components, the basis upon which the system is being evaluated, test results and similar information and other data as necessary for the code official to determine conformance to this code.

107.1.2.1 Evaluation service. The code official shall designate the evaluation service of an approved agency as the evaluation agency, and review such agency's evaluation report for adequacy and conformance to this code.

107.1.2.2 Follow-up inspection. Except where ready access is provided to installations, service equipment and accessories for complete inspection at the site without disassembly or dismantling, the code official shall conduct the in-plant inspections as frequently as necessary to ensure conformance to the approved evaluation report or shall designate an independent, approved inspection agency to conduct such inspections. The inspection agency shall furnish the code official with the follow-up inspection manual and a report of inspections upon request, and the installation shall have an identifying label permanently affixed to the system indicating that factory inspections have been performed.

107.1.2.3 Test and inspection records. Required test and inspection records shall be available to the code official at all times during the fabrication of the installation and the erection of the building; or such records as the code official designates shall be filed.

107.2 Testing. Installations shall be tested as required in this code and in accordance with Sections 107.2.1 through 107.2.3. Tests shall be made by the permit holder and observed by the code official.

107.2.1 New, altered, extended or repaired installations. New installations and parts of existing installations, which have been altered, extended, renovated or repaired, shall be tested as prescribed herein to disclose leaks and defects.

107.2.2 Apparatus, instruments, material and labor for tests. Apparatus, instruments, material and labor required for testing an installation or part thereof shall be furnished by the permit holder.

107.2.3 Reinspection and testing. Where any work or installation does not pass an initial test or inspection, the necessary corrections shall be made so as to achieve compliance with this code. The work or installation shall then be resubmitted to the code official for inspection and testing.

107.3 Approval. After the prescribed tests and inspections indicate that the work complies in all respects with this code, a notice of approval shall be issued by the code official.

107.4 Temporary connection. The code official shall have the authority to allow the temporary connection of an installation to the sources of energy for the purpose of testing the installation or for use under a temporary certificate of occupancy.

SECTION 108 (IFGC)
VIOLATIONS

108.1 Unlawful acts. It shall be unlawful for a person, firm or corporation to erect, construct, alter, repair, remove, demolish or utilize an installation, or cause same to be done, in conflict with or in violation of any of the provisions of this code.

108.2 Notice of violation. The code official shall serve a notice of violation or order to the person responsible for the erection, installation, alteration, extension, repair, removal or demolition of work in violation of the provisions of this code, or in violation of a detail statement or the approved construction documents thereunder, or in violation of a permit or certificate issued under the provisions of this code. Such order shall direct the discontinuance of the illegal action or condition and the abatement of the violation.

108.3 Prosecution of violation. If the notice of violation is not complied with promptly, the code official shall request the legal counsel of the jurisdiction to institute the appropriate proceeding at law or in equity to restrain, correct or abate such violation, or to require the removal or termination of the unlawful occupancy of the structure in violation of the provisions of this code or of the order or direction made pursuant thereto.

108.4 Violation penalties. Persons who shall violate a provision of this code, fail to comply with any of the requirements thereof or erect, install, alter or repair work in violation of the approved construction documents or directive of the code official, or of a permit or certificate issued under the provisions of this code, shall be guilty of a [SPECIFY OFFENSE], punishable by a fine of not more than [AMOUNT] dollars or by imprisonment not exceeding [NUMBER OF DAYS], or both such fine and imprisonment. Each day that a violation continues after due notice has been served shall be deemed a separate offense.

108.5 Stop work orders. Upon notice from the code official that work is being done contrary to the provisions of this code or in a dangerous or unsafe manner, such work shall immediately cease. Such notice shall be in writing and shall be given to the owner of the property, the owner's agent, or the person doing the work. The notice shall state the conditions under which work is authorized to resume. Where an emergency exists, the code official shall not be required to give a written notice prior to stopping the work. Any person who shall continue any work on the system after having been served with a stop work order, except such work as that person is directed to perform to remove a violation or unsafe condition, shall be liable for a fine of not less than [AMOUNT] dollars or more than [AMOUNT] dollars.

108.6 Abatement of violation. The imposition of the penalties herein prescribed shall not preclude the legal officer of the jurisdiction from instituting appropriate action to prevent unlawful construction, restrain, correct or abate a violation, prevent illegal occupancy of a building, structure or premises, or stop an illegal act, conduct, business or utilization of the installations on or about any premises.

108.7 Unsafe installations. An installation that is unsafe, constitutes a fire or health hazard, or is otherwise dangerous to human life, as regulated by this code, is hereby declared an unsafe installation. Use of an installation regulated by this code constituting a hazard to health, safety or welfare by reason of inadequate maintenance, dilapidation, fire hazard, disaster, damage or abandonment is hereby declared an unsafe use. Such unsafe installations are hereby declared to be a public nuisance and shall be abated by repair, rehabilitation, demolition or removal.

108.7.1 Authority to condemn installations. Whenever the code official determines that any installation, or portion thereof, regulated by this code has become hazardous to life, health or property, he or she shall order in writing that such installations either be removed or restored to a safe condition. A time limit for compliance with such order shall be specified in the written notice. A person shall not use or maintain a defective installation after receiving such notice.

When such installation is to be disconnected, written notice as prescribed in Section 108.2 shall be given. In cases of immediate danger to life or property, such disconnection shall be made immediately without such notice.

108.7.2 Authority to disconnect service utilities. The code official shall have the authority to require disconnection of utility service to the building, structure or system regulated by the technical codes in case of emergency where necessary to eliminate an immediate hazard to life or property. The code official shall notify the serving utility, and wherever possible, the owner and occupant of the building, structure or service system of the decision to disconnect prior to taking such action. If not notified prior to disconnection, the owner or occupant of the building, structure or service system shall be notified in writing, as soon as practicable thereafter.

108.7.3 Connection after order to disconnect. A person shall not make energy source connections to installations regulated by this code which have been disconnected or ordered to be disconnected by the code official, or the use of which has been ordered to be discontinued by the code official until the code official authorizes the reconnection and use of such installations.

When an installation is maintained in violation of this code, and in violation of a notice issued pursuant to the provisions of this section, the code official shall institute appropriate action to prevent, restrain, correct or abate the violation.

SECTION 109 (IFGC)
MEANS OF APPEAL

109.1 Application for appeal. A person shall have the right to appeal a decision of the code official to the board of appeals. An application for appeal shall be based on a claim that the true intent of this code or the rules legally adopted thereunder have been incorrectly interpreted, the provisions of this code do not fully apply or an equally good or better form of construction is proposed. The application shall be filed on a form obtained from the code official within 20 days after the notice was served.

109.2 Membership of board. The board of appeals shall consist of five members appointed by the chief appointing authority as follows: one for five years; one for four years; one for three years; one for two years and one for one year. Thereafter, each new member shall serve for five years or until a successor has been appointed.

109.2.1 Qualifications. The board of appeals shall consist of five individuals, one from each of the following professions or disciplines.

1. Registered design professional who is a registered architect; or a builder or superintendent of building construction with at least 10 years' experience, five of which shall have been in responsible charge of work.

2. Registered design professional with structural engineering or architectural experience.

3. Registered design professional with fuel gas and plumbing engineering experience; or a fuel gas contractor with at least 10 years' experience, five of which shall have been in responsible charge of work.

4. Registered design professional with electrical engineering experience; or an electrical contractor with at least 10 years' experience, five of which shall have been in responsible charge of work.

5. Registered design professional with fire protection engineering experience; or a fire protection contractor with at least 10 years' experience, five of which shall have been in responsible charge of work.

109.2.2 Alternate members. The chief appointing authority shall appoint two alternate members who shall be called by the board chairman to hear appeals during the absence or disqualification of a member. Alternate members shall possess the qualifications required for board membership and shall be appointed for five years, or until a successor has been appointed.

109.2.3 Chairman. The board shall annually select one of its members to serve as chairman.

109.2.4 Disqualification of member. A member shall not hear an appeal in which that member has a personal, professional or financial interest.

109.2.5 Secretary. The chief administrative officer shall designate a qualified clerk to serve as secretary to the board. The secretary shall file a detailed record of all proceedings in the office of the chief administrative officer.

109.2.6 Compensation of members. Compensation of members shall be determined by law.

109.3 Notice of meeting. The board shall meet upon notice from the chairman, within 10 days of the filing of an appeal, or at stated periodic meetings.

109.4 Open hearing. All hearings before the board shall be open to the public. The appellant, the appellant's representative, the code official and any person whose interests are affected shall be given an opportunity to be heard.

109.4.1 Procedure. The board shall adopt and make available to the public through the secretary procedures under which a hearing will be conducted. The procedures shall not require compliance with strict rules of evidence, but shall mandate that only relevant information be received.

109.5 Postponed hearing. When five members are not present to hear an appeal, either the appellant or the appellant's representative shall have the right to request a postponement of the hearing.

109.6 Board decision. The board shall modify or reverse the decision of the code official by a concurring vote of three members.

109.6.1 Resolution. The decision of the board shall be by resolution. Certified copies shall be furnished to the appellant and to the code official.

109.6.2 Administration. The code official shall take immediate action in accordance with the decision of the board.

109.7 Court review. Any person, whether or not a previous party to the appeal, shall have the right to apply to the appropriate court for a writ of certiorari to correct errors of law. Application for review shall be made in the manner and time required by law following the filing of the decision in the office of the chief administrative officer.

CHAPTER 2

DEFINITIONS

201.1 Scope. Unless otherwise expressly stated, the following words and terms shall, for the purposes of this code and standard, have the meanings indicated in this chapter.

201.2 Interchangeability. Words used in the present tense include the future; words in the masculine gender include the feminine and neuter; the singular number includes the plural and the plural, the singular.

201.3 Terms defined in other codes. Where terms are not defined in this code and are defined in the ICC *Electrical Code, International Building Code, International Fire Code, International Mechanical Code* or *International Plumbing Code*, such terms shall have meanings ascribed to them as in those codes.

201.4 Terms not defined. Where terms are not defined through the methods authorized by this section, such terms shall have ordinarily accepted meanings such as the context implies.

SECTION 202 (IFGC)
GENERAL DEFINITIONS

ACCESS (TO). That which enables a device, appliance or equipment to be reached by ready access or by a means that first requires the removal or movement of a panel, door or similar obstruction (see also "Ready access").

AIR CONDITIONER, GAS-FIRED. A gas-burning, automatically operated appliance for supplying cooled and/or dehumidified air or chilled liquid.

AIR CONDITIONING. The treatment of air so as to control simultaneously the temperature, humidity, cleanness and distribution of the air to meet the requirements of a conditioned space.

AIR, EXHAUST. Air being removed from any space or piece of equipment and conveyed directly to the atmosphere by means of openings or ducts.

AIR-HANDLING UNIT. A blower or fan used for the purpose of distributing supply air to a room, space or area.

AIR, MAKEUP. Air that is provided to replace air being exhausted.

ALTERATION. A change in a system that involves an extension, addition or change to the arrangement, type or purpose of the original installation.

ANODELESS RISER. A transition assembly in which plastic piping is installed and terminated above ground outside of a building.

APPLIANCE (EQUIPMENT). Any apparatus or equipment that utilizes gas as a fuel or raw material to produce light, heat, power, refrigeration or air conditioning.

APPLIANCE, FAN-ASSISTED COMBUSTION. An appliance equipped with an integral mechanical means to either draw or force products of combustion through the combustion chamber or heat exchanger.

APPLIANCE, AUTOMATICALLY CONTROLLED. Appliances equipped with an automatic burner ignition and safety shutoff device and other automatic devices which accomplish complete turn-on and shutoff of the gas to the main burner or burners, and graduate the gas supply to the burner or burners, but do not affect complete shutoff of the gas.

APPLIANCE TYPE.

Low-heat appliance (residential appliance). Any appliance in which the products of combustion at the point of entrance to the flue under normal operating conditions have a temperature of 1,000°F (538°C) or less.

Medium-heat appliance. Any appliance in which the products of combustion at the point of entrance to the flue under normal operating conditions have a temperature of more than 1,000°F (538°C), but not greater than 2,000°F (1093°C).

APPLIANCE, UNVENTED. An appliance designed or installed in such a manner that the products of combustion are not conveyed by a vent or chimney directly to the outside atmosphere.

APPLIANCE, VENTED. An appliance designed and installed in such a manner that all of the products of combustion are conveyed directly from the appliance to the outside atmosphere through an approved chimney or vent system.

APPROVED. Acceptable to the code official or other authority having jurisdiction.

APPROVED AGENCY. An established and recognized agency that is approved by the code official and regularly engaged in conducting tests or furnishing inspection services.

ATMOSPHERIC PRESSURE. The pressure of the weight of air and water vapor on the surface of the earth, approximately 14.7 pounds per square inch (psi) (101 kPa absolute) at sea level.

AUTOMATIC IGNITION. Ignition of gas at the burner(s) when the gas controlling device is turned on, including reignition if the flames on the burner(s) have been extinguished by means other than by the closing of the gas controlling device.

BAFFLE. An object placed in an appliance to change the direction of or retard the flow of air, air-gas mixtures or flue gases.

BAROMETRIC DRAFT REGULATOR. A balanced damper device attached to a chimney, vent connector, breeching or flue gas manifold to protect combustion equipment by controlling chimney draft. A double-acting barometric draft regulator is one whose balancing damper is free to move in either direction to protect combustion equipment from both excessive draft and backdraft.

BOILER, LOW-PRESSURE. A self-contained appliance for supplying steam or hot water.

Hot water heating boiler. A boiler in which no steam is generated, from which hot water is circulated for heating purposes and then returned to the boiler, and that operates at water pressures not exceeding 160 pounds per square inch gauge (psig) (1100 kPa gauge) and at water temperatures not exceeding 250°F (121°C) at or near the boiler outlet.

Hot water supply boiler. A boiler, completely filled with water, which furnishes hot water to be used externally to itself, and that operates at water pressures not exceeding 160 psig (1100 kPa gauge) and at water temperatures not exceeding 250°F (121°C) at or near the boiler outlet.

Steam heating boiler. A boiler in which steam is generated and that operates at a steam pressure not exceeding 15 psig (100 kPa gauge).

BRAZING. A metal-joining process wherein coalescence is produced by the use of a nonferrous filler metal having a melting point above 1,000°F (538°C), but lower than that of the base metal being joined. The filler material is distributed between the closely fitted surfaces of the joint by capillary action.

BROILER. A general term including salamanders, barbecues and other appliances cooking primarily by radiated heat, excepting toasters.

BTU. Abbreviation for British thermal unit, which is the quantity of heat required to raise the temperature of 1 pound (454 g) of water 1°F (0.56°C) (1 Btu = 1055 J).

BURNER. A device for the final conveyance of the gas, or a mixture of gas and air, to the combustion zone.

Induced-draft. A burner that depends on draft induced by a fan that is an integral part of the appliance and is located downstream from the burner.

Power. A burner in which gas, air or both are supplied at pressures exceeding, for gas, the line pressure, and for air, atmospheric pressure, with this added pressure being applied at the burner.

CHIMNEY. A primarily vertical structure containing one or more flues, for the purpose of carrying gaseous products of combustion and air from an appliance to the outside atmosphere.

Factory-built chimney. A listed and labeled chimney composed of factory-made components, assembled in the field in accordance with manufacturer's instructions and the conditions of the listing.

Masonry chimney. A field-constructed chimney composed of solid masonry units, bricks, stones or concrete.

Metal chimney. A field-constructed chimney of metal.

CLEARANCE. The minimum distance through air measured between the heat-producing surface of the mechanical appliance, device or equipment and the surface of the combustible material or assembly.

CLOTHES DRYER. An appliance used to dry wet laundry by means of heated air. Dryer classifications are as follows:

Type 1. Factory-built package, multiple production. Primarily used in family living environment. Usually the smallest unit physically and in function output.

Type 2. Factory-built package, multiple production. Used in business with direct intercourse of the function with the public. Not designed for use in individual family living environment.

CODE. These regulations, subsequent amendments thereto or any emergency rule or regulation that the administrative authority having jurisdiction has lawfully adopted.

CODE OFFICIAL. The officer or other designated authority charged with the administration and enforcement of this code, or a duly authorized representative.

COMBUSTION. In the context of this code, refers to the rapid oxidation of fuel accompanied by the production of heat or heat and light.

COMBUSTION AIR. Air necessary for complete combustion of a fuel, including theoretical air and excess air.

COMBUSTION CHAMBER. The portion of an appliance within which combustion occurs.

COMBUSTION PRODUCTS. Constituents resulting from the combustion of a fuel with the oxygen of the air, including inert gases, but excluding excess air.

CONCEALED LOCATION. A location that cannot be accessed without damaging permanent parts of the building structure or finish surface. Spaces above, below or behind readily removable panels or doors shall not be considered as concealed.

CONCEALED PIPING. Piping that is located in a concealed location (see "Concealed location").

CONDENSATE. The liquid that condenses from a gas (including flue gas) caused by a reduction in temperature or increase in pressure.

CONNECTOR, APPLIANCE (Fuel). Rigid metallic pipe and fittings, semirigid metallic tubing and fittings or a listed and labeled device that connects an appliance to the gas piping system.

CONNECTOR, CHIMNEY OR VENT. The pipe that connects an appliance to a chimney or vent.

CONSTRUCTION DOCUMENTS. All of the written, graphic and pictorial documents prepared or assembled for describing the design, location and physical characteristics of the elements of the project necessary for obtaining a mechanical permit.

CONTROL. A manual or automatic device designed to regulate the gas, air, water or electrical supply to, or operation of, a mechanical system.

CONVERSION BURNER. A unit consisting of a burner and its controls for installation in an appliance originally utilizing another fuel.

COUNTER APPLIANCES. Appliances such as coffee brewers and coffee urns and any appurtenant water-heating equipment, food and dish warmers, hot plates, griddles, waffle bakers and other appliances designed for installation on or in a counter.

CUBIC FOOT. The amount of gas that occupies 1 cubic foot (0.02832 m³) when at a temperature of 60°F (16°C), saturated with water vapor and under a pressure equivalent to that of 30 inches of mercury (101 kPa).

DAMPER. A manually or automatically controlled device to regulate draft or the rate of flow of air or combustion gases.

DECORATIVE APPLIANCE, VENTED. A vented appliance wherein the primary function lies in the aesthetic effect of the flames.

DECORATIVE APPLIANCES FOR INSTALLATION IN VENTED FIREPLACES. A vented appliance designed for installation within the fire chamber of a vented fireplace, wherein the primary function lies in the aesthetic effect of the flames.

DEMAND. The maximum amount of gas input required per unit of time, usually expressed in cubic feet per hour, or Btu/h (1 Btu/h = 0.2931 W).

DESIGN FLOOD ELEVATION. The elevation of the "design flood," including wave height, relative to the datum specified on the community's legally designated flood hazard map.

DILUTION AIR. Air that is introduced into a draft hood and is mixed with the flue gases.

DIRECT-VENT APPLIANCES. Appliances that are constructed and installed so that all air for combustion is derived directly from the outside atmosphere and all flue gases are discharged directly to the outside atmosphere.

DRAFT. The pressure difference existing between the equipment or any component part and the atmosphere, that causes a continuous flow of air and products of combustion through the gas passages of the appliance to the atmosphere.

Mechanical or induced draft. The pressure difference created by the action of a fan, blower or ejector, that is located between the appliance and the chimney or vent termination.

Natural draft. The pressure difference created by a vent or chimney because of its height, and the temperature difference between the flue gases and the atmosphere.

DRAFT HOOD. A nonadjustable device built into an appliance, or made as part of the vent connector from an appliance, that is designed to (1) provide for ready escape of the flue gases from the appliance in the event of no draft, backdraft or stoppage beyond the draft hood, (2) prevent a backdraft from entering the appliance, and (3) neutralize the effect of stack action of the chimney or gas vent upon operation of the appliance.

DRAFT REGULATOR. A device that functions to maintain a desired draft in the appliance by automatically reducing the draft to the desired value.

DRIP. The container placed at a low point in a system of piping to collect condensate and from which the condensate is removable.

DRY GAS. A gas having a moisture and hydrocarbon dew point below any normal temperature to which the gas piping is exposed.

DUCT FURNACE. A warm-air furnace normally installed in an air distribution duct to supply warm air for heating. This definition shall apply only to a warm-air heating appliance that depends for air circulation on a blower not furnished as part of the furnace.

DUCT SYSTEM. A continuous passageway for the transmission of air that, in addition to ducts, includes duct fittings, dampers, plenums, fans and accessory air-handling equipment.

DWELLING UNIT. A single unit providing complete, independent living facilities for one or more persons, including permanent provisions for living, sleeping, eating, cooking and sanitation.

EQUIPMENT. See "Appliance."

FIREPLACE. A fire chamber and hearth constructed of noncombustible material for use with solid fuels and provided with a chimney.

Masonry fireplace. A hearth and fire chamber of solid masonry units such as bricks, stones, listed masonry units or reinforced concrete, provided with a suitable chimney.

Factory-built fireplace. A fireplace composed of listed factory-built components assembled in accordance with the terms of listing to form the completed fireplace.

FIRING VALVE. A valve of the plug and barrel type designed for use with gas, and equipped with a lever handle for manual operation and a dial to indicate the percentage of opening.

FLAME SAFEGUARD. A device that will automatically shut off the fuel supply to a main burner or group of burners when the means of ignition of such burners becomes inoperative, and when flame failure occurs on the burner or group of burners.

FLOOD HAZARD AREA. The greater of the following two areas:

1. The area within a floodplain subject to a 1 percent or greater chance of flooding in any given year.
2. This area designated as a flood hazard area on a community's flood hazard map, or otherwise legally designated.

FLOOR FURNACE. A completely self-contained furnace suspended from the floor of the space being heated, taking air for combustion from outside such space and with means for observing flames and lighting the appliance from such space.

Gravity type. A floor furnace depending primarily upon circulation of air by gravity. This classification shall also include floor furnaces equipped with booster-type fans which do not materially restrict free circulation of air by gravity flow when such fans are not in operation.

Fan type. A floor furnace equipped with a fan which provides the primary means for circulating air.

FLUE, APPLIANCE. The passage(s) within an appliance through which combustion products pass from the combustion chamber of the appliance to the draft hood inlet opening on an appliance equipped with a draft hood or to the outlet of the appliance on an appliance not equipped with a draft hood.

FLUE COLLAR. That portion of an appliance designed for the attachment of a draft hood, vent connector or venting system.

FLUE GASES. Products of combustion plus excess air in appliance flues or heat exchangers.

FLUE LINER (LINING). A system or material used to form the inside surface of a flue in a chimney or vent, for the purpose of protecting the surrounding structure from the effects of combustion products and for conveying combustion products without leakage to the atmosphere.

FUEL GAS. A natural gas, manufactured gas, liquefied petroleum gas or mixtures of these gases.

FUEL GAS UTILIZATION EQUIPMENT. See "Appliance."

FURNACE. A completely self-contained heating unit that is designed to supply heated air to spaces remote from or adjacent to the appliance location.

FURNACE, CENTRAL. A self-contained appliance for heating air by transfer of heat of combustion through metal to the air, and designed to supply heated air through ducts to spaces remote from or adjacent to the appliance location.

> **Downflow furnace.** A furnace designed with airflow discharge vertically downward at or near the bottom of the furnace.

> **Forced air furnace with cooling unit.** A single-package unit, consisting of a gas-fired forced-air furnace of one of the types listed below combined with an electrically or fuel gas-powered summer air-conditioning system, contained in a common casing.

> **Forced-air type.** A central furnace equipped with a fan or blower which provides the primary means for circulation of air.

> **Gravity furnace with booster fan.** A furnace equipped with a booster fan that does not materially restrict free circulation of air by gravity flow when the fan is not in operation.

> **Gravity type.** A central furnace depending primarily on circulation of air by gravity.

> **Horizontal forced-air type.** A furnace with airflow through the appliance essentially in a horizontal path.

> **Multiple-position furnace.** A furnace designed so that it can be installed with the airflow discharge in the upflow, horizontal or downflow direction.

> **Upflow furnace.** A furnace designed with airflow discharge vertically upward at or near the top of the furnace. This classification includes "highboy" furnaces with the blower mounted below the heating element and "lowboy" furnaces with the blower mounted beside the heating element.

FURNACE, ENCLOSED. A specific heating, or heating and ventilating, furnace incorporating an integral total enclosure and using only outside air for combustion.

FURNACE PLENUM. An air compartment or chamber to which one or more ducts are connected and which forms part of an air distribution system.

GAS CONVENIENCE OUTLET. A permanently mounted, manually operated device that provides the means for connecting an appliance to, and disconnecting an appliance from, the supply piping. The device includes an integral, manually operated valve with a nondisplaceable valve member and is designed so that disconnection of an appliance only occurs when the manually operated valve is in the closed position.

GASEOUS HYDROGEN SYSTEM. See Section 702.1.

GAS PIPING. An installation of pipe, valves or fittings installed on a premises or in a building and utilized to convey fuel gas.

GAS UTILIZATION EQUIPMENT. An appliance that utilizes gas as a fuel or raw material or both.

HAZARDOUS LOCATION. Any location considered to be a fire hazard for flammable vapors, dust, combustible fibers or other highly combustible substances. The location is not necessarily categorized in the building code as a high-hazard group classification.

HOUSE PIPING. See "Piping system."

HYDROGEN CUT-OFF ROOM. See Section 702.1.

HYDROGEN GENERATING APPLIANCE. See Section 702.1.

IGNITION PILOT. A pilot that operates during the lighting cycle and discontinues during main burner operation.

IGNITION SOURCE. A flame, spark or hot surface capable of igniting flammable vapors or fumes. Such sources include appliance burners, burner ignitors, and electrical switching devices.

INCINERATOR. An appliance used to reduce combustible refuse material to ashes and which is manufactured, sold and installed as a complete unit.

INDUSTRIAL AIR HEATERS, DIRECT-FIRED NONRECIRCULATING. A heater in which all the products of combustion generated by the burners are released into the air stream being heated. The purpose of the heater is to offset building heat loss by heating only outdoor air.

INDUSTRIAL AIR HEATERS, DIRECT-FIRED RECIRCULATING. A heater in which all the products of combustion generated by the burners are released into the air stream being heated. The purpose of the heater is to offset building heat loss by heating outdoor air, and, if applicable, indoor air.

INFRARED RADIANT HEATER. A heater that directs a substantial amount of its energy output in the form of infrared radiant energy into the area to be heated. Such heaters are of either the vented or unvented type.

JOINT, FLANGED. A joint made by bolting together a pair of flanged ends.

JOINT, FLARED. A metal-to-metal compression joint in which a conical spread is made on the end of a tube that is compressed by a flare nut against a mating flare.

JOINT, MECHANICAL. A general form of gas-tight joints obtained by the joining of metal parts through a positive-holding mechanical construction, such as flanged joint, threaded joint, flared joint or compression joint.

JOINT, PLASTIC ADHESIVE. A joint made in thermoset plastic piping by the use of an adhesive substance which forms a continuous bond between the mating surfaces without dissolving either one of them.

JOINT, PLASTIC HEAT FUSION. A joint made in thermoplastic piping by heating the parts sufficiently to permit fusion of the materials when the parts are pressed together.

JOINT, WELDED. A gas-tight joint obtained by the joining of metal parts in molten state.

LABELED. Devices, equipment, appliances or materials to which have been affixed a label, seal, symbol or other identifying mark of a nationally recognized testing laboratory, inspection agency or other organization concerned with product evaluation that maintains periodic inspection of the production of the above-labeled items and by whose label the manufacturer attests to compliance with applicable nationally recognized standards.

LIMIT CONTROL. A device responsive to changes in pressure, temperature or level for turning on, shutting off or throttling the gas supply to an appliance.

LIQUEFIED PETROLEUM GAS or LPG (LP-GAS). Liquefied petroleum gas composed predominately of propane, propylene, butanes or butylenes, or mixtures thereof that is gaseous under normal atmospheric conditions, but is capable of being liquefied under moderate pressure at normal temperatures.

LISTED. Equipment, appliances or materials included in a list published by a nationally recognized testing laboratory, inspection agency or other organization concerned with product evaluation that maintains periodic inspection of production of listed equipment, appliances or materials, and whose listing states either that the equipment, appliance or material meets nationally recognized standards or has been tested and found suitable for use in a specified manner. The means for identifying listed equipment, appliances or materials may vary for each testing laboratory, inspection agency or other organization concerned with product evaluation, some of which do not recognize equipment, appliances or materials as listed unless they are also labeled. The authority having jurisdiction shall utilize the system employed by the listing organization to identify a listed product.

LIVING SPACE. Space within a dwelling unit utilized for living, sleeping, eating, cooking, bathing, washing and sanitation purposes.

LOG LIGHTER. A manually operated solid fuel ignition appliance for installation in a vented solid fuel-burning fireplace.

LUBRICATED PLUG-TYPE VALVE. A valve of the plug and barrel type provided with means for maintaining a lubricant between the bearing surfaces.

MAIN BURNER. A device or group of devices essentially forming an integral unit for the final conveyance of gas or a mixture of gas and air to the combustion zone, and on which combustion takes place to accomplish the function for which the appliance is designed.

METER. The instrument installed to measure the volume of gas delivered through it.

MODULATING. Modulating or throttling is the action of a control from its maximum to minimum position in either predetermined steps or increments of movement as caused by its actuating medium.

OCCUPANCY. The purpose for which a building, or portion thereof, is utilized or occupied.

OFFSET (VENT). A combination of approved bends that makes two changes in direction bringing one section of the vent out of line but into a line parallel with the other section.

ORIFICE. The opening in a cap, spud or other device whereby the flow of gas is limited and through which the gas is discharged to the burner.

OUTLET. A threaded connection or bolted flange in a pipe system to which a gas-burning appliance is attached.

OXYGEN DEPLETION SAFETY SHUTOFF SYSTEM (ODS). A system designed to act to shut off the gas supply to the main and pilot burners if the oxygen in the surrounding atmosphere is reduced below a predetermined level.

PILOT. A small flame that is utilized to ignite the gas at the main burner or burners.

PIPING. Where used in this code, "piping" refers to either pipe or tubing, or both.

> **Pipe.** A rigid conduit of iron, steel, copper, brass or plastic.

> **Tubing.** Semirigid conduit of copper, aluminum, plastic or steel.

PIPING SYSTEM. All fuel piping, valves and fittings from the outlet of the point of delivery to the outlets of the equipment shutoff valves.

PLASTIC, THERMOPLASTIC. A plastic that is capable of being repeatedly softened by increase of temperature and hardened by decrease of temperature.

POINT OF DELIVERY. For natural gas systems, the point of delivery is the outlet of the service meter assembly or the outlet of the service regulator or service shutoff valve where a meter is not provided. Where a valve is provided at the outlet of the service meter assembly, such valve shall be considered to be downstream of the point of delivery. For undiluted liquefied petroleum gas systems, the point of delivery shall be considered to be the outlet of the first regulator that reduces pressure to 2 psig (13.8 kPag) or less.

PORTABLE FUEL CELL APPLIANCE. A fuel cell generator of electricity, which is not fixed in place. A portable fuel cell appliance utilizes a cord and plug connection to a grid-isolated load and has an integral fuel supply.

PRESSURE DROP. The loss in pressure due to friction or obstruction in pipes, valves, fittings, regulators and burners.

PRESSURE TEST. An operation performed to verify the gas-tight integrity of gas piping following its installation or modification.

PURGE. To free a gas conduit of air or gas, or a mixture of gas and air.

QUICK-DISCONNECT DEVICE. A hand-operated device that provides a means for connecting and disconnecting an appliance or an appliance connector to a gas supply and that is equipped with an automatic means to shut off the gas supply when the device is disconnected.

READY ACCESS (TO). That which enables a device, appliance or equipment to be directly reached, without requiring the removal or movement of any panel, door or similar obstruction (see "Access").

REGISTERED DESIGN PROFESSIONAL. An individual who is registered or licensed to practice their respective design profession as defined by the statutory requirements of the professional registration laws of the state or jurisdiction in which the project is to be constructed.

REGULATOR. A device for controlling and maintaining a uniform supply pressure, either pounds-to-inches water column (MP regulator) or inches-to-inches water column (appliance regulator).

REGULATOR, GAS APPLIANCE. A pressure regulator for controlling pressure to the manifold of equipment. Types of appliance regulators are as follows:

Adjustable.

1. Spring type, limited adjustment. A regulator in which the regulating force acting upon the diaphragm is derived principally from a spring, the loading of which is adjustable over a range of not more than 15 percent of the outlet pressure at the midpoint of the adjustment range.

2. Spring type, standard adjustment. A regulator in which the regulating force acting upon the diaphragm is derived principally from a spring, the loading of which is adjustable. The adjustment means shall be concealed.

Multistage. A regulator for use with a single gas whose adjustment means is capable of being positioned manually or automatically to two or more predetermined outlet pressure settings. Each of these settings shall be adjustable or nonadjustable. The regulator may modulate outlet pressures automatically between its maximum and minimum predetermined outlet pressure settings.

Nonadjustable.

1. Spring type, nonadjustable. A regulator in which the regulating force acting upon the diaphragm is derived principally from a spring, the loading of which is not field adjustable.

2. Weight type. A regulator in which the regulating force acting upon the diaphragm is derived from a weight or combination of weights.

REGULATOR, LINE GAS PRESSURE. A device placed in a gas line between the service pressure regulator and the equipment for controlling, maintaining or reducing the pressure in that portion of the piping system downstream of the device.

REGULATOR, MEDIUM-PRESSURE (MP Regulator). A line pressure regulator that reduces gas pressure from the range of greater than 0.5 psig (3.4 kPa) and less than or equal to 5 psig (34.5 kPa) to a lower pressure.

REGULATOR, PRESSURE. A device placed in a gas line for reducing, controlling and maintaining the pressure in that portion of the piping system downstream of the device.

REGULATOR, SERVICE PRESSURE. A device installed by the serving gas supplier to reduce and limit the service line pressure to delivery pressure.

RELIEF OPENING. The opening provided in a draft hood to permit the ready escape to the atmosphere of the flue products from the draft hood in the event of no draft, back draft, or stoppage beyond the draft hood, and to permit air into the draft hood in the event of a strong chimney updraft.

RELIEF VALVE (DEVICE). A safety valve designed to forestall the development of a dangerous condition by relieving either pressure, temperature or vacuum in the hot water supply system.

RELIEF VALVE, PRESSURE. An automatic valve that opens and closes a relief vent, depending on whether the pressure is above or below a predetermined value.

RELIEF VALVE, TEMPERATURE.

Reseating or self-closing type. An automatic valve that opens and closes a relief vent, depending on whether the temperature is above or below a predetermined value.

Manual reset type. A valve that automatically opens a relief vent at a predetermined temperature and that must be manually returned to the closed position.

RELIEF VALVE, VACUUM. A valve that automatically opens and closes a vent for relieving a vacuum within the hot water supply system, depending on whether the vacuum is above or below a predetermined value.

RISER, GAS. A vertical pipe supplying fuel gas.

ROOM HEATER, UNVENTED. See "Unvented room heater."

ROOM HEATER, VENTED. A free-standing heating unit used for direct heating of the space in and adjacent to that in which the unit is located (see also "Vented room heater").

ROOM LARGE IN COMPARISON WITH SIZE OF EQUIPMENT. Rooms having a volume equal to at least 12 times the total volume of a furnace or air-conditioning appliance and at least 16 times the total volume of a boiler. Total volume of the appliance is determined from exterior dimensions and is to include fan compartments and burner vestibules, when used. When the actual ceiling height of a room is greater than 8 feet (2438 mm), the volume of the room is figured on the basis of a ceiling height of 8 feet (2438 mm).

SAFETY SHUTOFF DEVICE. See "Flame safeguard."

SHAFT. An enclosed space extending through one or more stories of a building, connecting vertical openings in successive floors, or floors and the roof.

SLEEPING UNIT. A room or space in which people sleep, which can also include permanent provisions for living, eating and either sanitation or kitchen facilities, but not both. Such rooms and spaces that are also part of a dwelling unit are not sleeping units.

SPECIFIC GRAVITY. As applied to gas, specific gravity is the ratio of the weight of a given volume to that of the same volume of air, both measured under the same condition.

STATIONARY FUEL CELL POWER PLANT. A self-contained package or factory-matched packages which constitute an automatically operated assembly of integrated systems for generating electrical energy and recoverable thermal energy that is permanently connected and fixed in place.

THERMOSTAT.

Electric switch type. A device that senses changes in temperature and controls electrically, by means of separate components, the flow of gas to the burner(s) to maintain selected temperatures.

Integral gas valve type. An automatic device, actuated by temperature changes, designed to control the gas supply to the burner(s) in order to maintain temperatures between predetermined limits, and in which the thermal actuating element is an integral part of the device.

1. Graduating thermostat. A thermostat in which the motion of the valve is approximately in direct proportion to the effective motion of the thermal element induced by temperature change.

2. Snap-acting thermostat. A thermostat in which the thermostatic valve travels instantly from the closed to the open position, and vice versa.

TRANSITION FITTINGS, PLASTIC TO STEEL. An adapter for joining plastic pipe to steel pipe. The purpose of this fitting is to provide a permanent, pressure-tight connection between two materials which cannot be joined directly one to another.

UNIT HEATER.

High-static pressure type. A self-contained, automatically controlled, vented appliance having integral means for circulation of air against 0.2 inch (15 mm H_2O) or greater static pressure. Such appliance is equipped with provisions for attaching an outlet air duct and, where the appliance is for indoor installation remote from the space to be heated, is also equipped with provisions for attaching an inlet air duct.

Low-static pressure type. A self-contained, automatically controlled, vented appliance, intended for installation in the space to be heated without the use of ducts, having integral means for circulation of air. Such units are allowed to be equipped with louvers or face extensions made in accordance with the manufacturer's specifications.

UNLISTED BOILER. A boiler not listed by a nationally recognized testing agency.

UNVENTED ROOM HEATER. An unvented heating appliance designed for stationary installation and utilized to provide comfort heating. Such appliances provide radiant heat or convection heat by gravity or fan circulation directly from the heater and do not utilize ducts.

VALVE. A device used in piping to control the gas supply to any section of a system of piping or to an appliance.

Automatic. An automatic or semiautomatic device consisting essentially of a valve and operator that control the gas supply to the burner(s) during operation of an appliance. The operator shall be actuated by application of gas pressure on a flexible diaphragm, by electrical means, by mechanical means, or by other approved means.

Automatic gas shutoff. A valve used in conjunction with an automatic gas shutoff device to shut off the gas supply to a water-heating system. It shall be constructed integrally with the gas shutoff device or shall be a separate assembly.

Equipment shutoff. A valve located in the piping system, used to isolate individual equipment for purposes such as service or replacement.

Individual main burner. A valve that controls the gas supply to an individual main burner.

Main burner control. A valve that controls the gas supply to the main burner manifold.

Manual main gas-control. A manually operated valve in the gas line for the purpose of completely turning on or shutting off the gas supply to the appliance, except to pilot or pilots that are provided with independent shutoff.

Manual reset. An automatic shutoff valve installed in the gas supply piping and set to shut off when unsafe conditions occur. The device remains closed until manually reopened.

Service shutoff. A valve, installed by the serving gas supplier between the service meter or source of supply and the customer piping system, to shut off the entire piping system.

VENT. A pipe or other conduit composed of factory-made components, containing a passageway for conveying combustion products and air to the atmosphere, listed and labeled for use with a specific type or class of appliance.

Special gas vent. A vent listed and labeled for use with listed Category II, III and IV appliances.

Type B vent. A vent listed and labeled for use with appliances with draft hoods and other Category I appliances that are listed for use with Type B vents.

Type BW vent. A vent listed and labeled for use with wall furnaces.

Type L vent. A vent listed and labeled for use with appliances that are listed for use with Type L or Type B vents.

VENT CONNECTOR. See "Connector."

VENT GASES. Products of combustion from appliances plus excess air plus dilution air in the vent connector, gas vent or chimney above the draft hood or draft regulator.

VENT PIPING

Breather. Piping run from a pressure-regulating device to the outdoors, designed to provide a reference to atmospheric pressure. If the device incorporates an integral pressure relief mechanism, a breather vent can also serve as a relief vent.

Relief. Piping run from a pressure-regulating or pressure-limiting device to the outdoors, designed to provide for the safe venting of gas in the event of excessive pressure in the gas piping system.

VENTED APPLIANCE CATEGORIES. Appliances that are categorized for the purpose of vent selection are classified into the following four categories:

Category I. An appliance that operates with a nonpositive vent static pressure and with a vent gas temperature that avoids excessive condensate production in the vent.

Category II. An appliance that operates with a nonpositive vent static pressure and with a vent gas temperature that is capable of causing excessive condensate production in the vent.

Category III. An appliance that operates with a positive vent static pressure and with a vent gas temperature that avoids excessive condensate production in the vent.

Category IV. An appliance that operates with a positive vent static pressure and with a vent gas temperature that is capable of causing excessive condensate production in the vent.

VENTED ROOM HEATER. A vented self-contained, free-standing, nonrecessed appliance for furnishing warm air to the space in which it is installed, directly from the heater without duct connections.

VENTED WALL FURNACE. A self-contained vented appliance complete with grilles or equivalent, designed for incorporation in or permanent attachment to the structure of a building, mobile home or travel trailer, and furnishing heated air circulated by gravity or by a fan directly into the space to be heated through openings in the casing. This definition shall exclude floor furnaces, unit heaters and central furnaces as herein defined.

VENTING SYSTEM. A continuous open passageway from the flue collar or draft hood of an appliance to the outside atmosphere for the purpose of removing flue or vent gases. A venting system is usually composed of a vent or a chimney and vent connector, if used, assembled to form the open passageway.

Mechanical draft venting system. A venting system designed to remove flue or vent gases by mechanical means, that consists of an induced draft portion under nonpositive static pressure or a forced draft portion under positive static pressure.

Forced-draft venting system. A portion of a venting system using a fan or other mechanical means to cause the removal of flue or vent gases under positive static vent pressure.

Induced draft venting system. A portion of a venting system using a fan or other mechanical means to cause the removal of flue or vent gases under nonpositive static vent pressure.

Natural draft venting system. A venting system designed to remove flue or vent gases under nonpositive static vent pressure entirely by natural draft.

WALL HEATER, UNVENTED-TYPE. A room heater of the type designed for insertion in or attachment to a wall or partition. Such heater does not incorporate concealed venting arrangements in its construction and discharges all products of combustion through the front into the room being heated.

WATER HEATER. Any heating appliance or equipment that heats potable water and supplies such water to the potable hot water distribution system.

CHAPTER 3
GENERAL REGULATIONS

301.1 Scope. This chapter shall govern the approval and installation of all equipment and appliances that comprise parts of the installations regulated by this code in accordance with Section 101.2.

301.1.1 Other fuels. The requirements for combustion and dilution air for gas-fired appliances shall be governed by Section 304. The requirements for combustion and dilution air for appliances operating with fuels other than fuel gas shall be regulated by the *International Mechanical Code*.

301.2 Energy utilization. Heating, ventilating and air-conditioning systems of all structures shall be designed and installed for efficient utilization of energy in accordance with the *International Energy Conservation Code*.

301.3 Listed and labeled. Appliances regulated by this code shall be listed and labeled for the application in which they are used unless otherwise approved in accordance with Section 105. The approval of unlisted appliances in accordance with Section 105 shall be based upon approved engineering evaluation.

301.4 Labeling. Labeling shall be in accordance with the procedures set forth in Sections 301.4.1 through 301.4.2.3.

301.4.1 Testing. An approved agency shall test a representative sample of the appliances being labeled to the relevant standard or standards. The approved agency shall maintain a record of all of the tests performed. The record shall provide sufficient detail to verify compliance with the test standard.

301.4.2 Inspection and identification. The approved agency shall periodically perform an inspection, which shall be in-plant if necessary, of the appliances to be labeled. The inspection shall verify that the labeled appliances are representative of the appliances tested.

301.4.2.1 Independent. The agency to be approved shall be objective and competent. To confirm its objectivity, the agency shall disclose all possible conflicts of interest.

301.4.2.2 Equipment. An approved agency shall have adequate equipment to perform all required tests. The equipment shall be periodically calibrated.

301.4.2.3 Personnel. An approved agency shall employ experienced personnel educated in conducting, supervising and evaluating tests.

301.5 Label information. A permanent factory-applied nameplate(s) shall be affixed to appliances on which shall appear in legible lettering, the manufacturer's name or trademark, the model number, serial number and, for listed appliances, the seal or mark of the testing agency. A label shall also include the hourly rating in British thermal units per hour (Btu/h) (W); the type of fuel approved for use with the appliance; and the minimum clearance requirements.

301.6 Plumbing connections. Potable water supply and building drainage system connections to appliances regulated by this code shall be in accordance with the *International Plumbing Code*.

301.7 Fuel types. Appliances shall be designed for use with the type of fuel gas that will be supplied to them.

301.7.1 Appliance fuel conversion. Appliances shall not be converted to utilize a different fuel gas except where complete instructions for such conversion are provided in the installation instructions, by the serving gas supplier or by the appliance manufacturer.

301.8 Vibration isolation. Where means for isolation of vibration of an appliance is installed, an approved means for support and restraint of that appliance shall be provided.

301.9 Repair. Defective material or parts shall be replaced or repaired in such a manner so as to preserve the original approval or listing.

301.10 Wind resistance. Appliances and supports that are exposed to wind shall be designed and installed to resist the wind pressures determined in accordance with the *International Building Code*.

301.11 Flood hazard. For structures located in flood hazard areas, the appliance, equipment and system installations regulated by this code shall be located at or above the design flood elevation and shall comply with the flood-resistant construction requirements of the *International Building Code*.

Exception: The appliance, equipment and system installations regulated by this code are permitted to be located below the design flood elevation provided that they are designed and installed to prevent water from entering or accumulating within the components and to resist hydrostatic and hydrodynamic loads and stresses, including the effects of buoyancy, during the occurrence of flooding to the design flood elevation and shall comply with the flood-resistant construction requirements of the *International Building Code*.

301.12 Seismic resistance. When earthquake loads are applicable in accordance with the *International Building Code*, the supports shall be designed and installed for the seismic forces in accordance with that code.

301.13 Ducts. All ducts required for the installation of systems regulated by this code shall be designed and installed in accordance with the *International Mechanical Code*.

301.14 Rodentproofing. Buildings or structures and the walls enclosing habitable or occupiable rooms and spaces in which persons live, sleep or work, or in which feed, food or foodstuffs are stored, prepared, processed, served or sold, shall be constructed to protect against rodents in accordance with the *International Building Code*.

301.15 Prohibited location. The appliances, equipment and systems regulated by this code shall not be located in an elevator shaft.

SECTION 302 (IFGC)
STRUCTURAL SAFETY

[B] 302.1 Structural safety. The building shall not be weakened by the installation of any gas piping. In the process of installing or repairing any gas piping, the finished floors, walls, ceilings, tile work or any other part of the building or premises which is required to be changed or replaced shall be left in a safe structural condition in accordance with the requirements of the *International Building Code.*

[B] 302.2 Penetrations of floor/ceiling assemblies and fire-resistance-rated assemblies. Penetrations of floor/ceiling assemblies and assemblies required to have a fire-resistance rating shall be protected in accordance with the *International Building Code.*

[B] 302.3 Cutting, notching and boring in wood members. The cutting, notching and boring of wood members shall comply with Sections 302.3.1 through 302.3.4.

[B] 302.3.1 Engineered wood products. Cuts, notches and holes bored in trusses, structural composite lumber, structural glued-laminated members and I-joists are prohibited except where permitted by the manufacturer's recommendations or where the effects of such alterations are specifically considered in the design of the member by a registered design professional.

[B] 302.3.2 Joist notching and boring. Notching at the ends of joists shall not exceed one-fourth the joist depth. Holes bored in joists shall not be within 2 inches (51 mm) of the top and bottom of the joist and their diameter shall not exceed one-third the depth of the member. Notches in the top or bottom of the joist shall not exceed one-sixth the depth and shall not be located in the middle one-third of the span.

[B] 302.3.3 Stud cutting and notching. In exterior walls and bearing partitions, any wood stud is permitted to be cut or notched to a depth not exceeding 25 percent of its width. Cutting or notching of studs to a depth not greater than 40 percent of the width of the stud is permitted in nonload-bearing partitions supporting no loads other than the weight of the partition.

[B] 302.3.4 Bored holes. A hole not greater in diameter than 40 percent of the stud depth is permitted to be bored in any wood stud. Bored holes not greater than 60 percent of the depth of the stud are permitted in nonload-bearing partitions or in any wall where each bored stud is doubled, provided not more than two such successive doubled studs are so bored. In no case shall the edge of the bored hole be nearer than $5/_8$ inch (15.9 mm) to the edge of the stud. Bored holes shall not be located at the same section of a stud as a cut or notch.

[B] 302.4 Alterations to trusses. Truss members and components shall not be cut, drilled, notched, spliced or otherwise altered in any way without the written concurrence and approval of a registered design professional. Alterations resulting in the addition of loads to any member (e.g., HVAC equipment, water heaters) shall not be permitted without verification that the truss is capable of supporting such additional loading.

[B] 302.5 Cutting, notching and boring holes in structural steel framing. The cutting, notching and boring of holes in structural steel framing members shall be as prescribed by the registered design professional.

[B] 302.6 Cutting, notching and boring holes in cold-formed steel framing. Flanges and lips of load-bearing, cold-formed steel framing members shall not be cut or notched. Holes in webs of load-bearing, cold-formed steel framing members shall be permitted along the centerline of the web of the framing member and shall not exceed the dimensional limitations, penetration spacing or minimum hole edge distance as prescribed by the registered design professional. Cutting, notching and boring holes of steel floor/roof decking shall be as prescribed by the registered design professional.

[B] 302.7 Cutting, notching and boring holes in nonstructural cold-formed steel wall framing. Flanges and lips of nonstructural cold-formed steel wall studs shall be permitted along the centerline of the web of the framing member, shall not exceed $1^1/_2$ inches (38 mm) in width or 4 inches (102 mm) in length, and the holes shall not be spaced less than 24 inches (610 mm) center to center from another hole or less than 10 inches (254 mm) from the bearing end.

SECTION 303 (IFGC)
APPLIANCE LOCATION

303.1 General. Appliances shall be located as required by this section, specific requirements elsewhere in this code and the conditions of the equipment and appliance listing.

303.2 Hazardous locations. Appliances shall not be located in a hazardous location unless listed and approved for the specific installation.

303.3 Prohibited locations. Appliances shall not be located in sleeping rooms, bathrooms, toilet rooms, storage closets or surgical rooms, or in a space that opens only into such rooms or spaces, except where the installation complies with one of the following:

1. The appliance is a direct-vent appliance installed in accordance with the conditions of the listing and the manufacturer's instructions.

2. Vented room heaters, wall furnaces, vented decorative appliances, vented gas fireplaces, vented gas fireplace heaters and decorative appliances for installation in vented solid fuel-burning fireplaces are installed in rooms that meet the required volume criteria of Section 304.5.

3. A single wall-mounted unvented room heater is installed in a bathroom and such unvented room heater is equipped as specified in Section 621.6 and has an input rating not greater than 6,000 Btu/h (1.76 kW). The bathroom shall meet the required volume criteria of Section 304.5.

4. A single wall-mounted unvented room heater is installed in a bedroom and such unvented room heater is equipped as specified in Section 621.6 and has an input rating not greater than 10,000 Btu/h (2.93 kW). The bedroom shall meet the required volume criteria of Section 304.5.

5. The appliance is installed in a room or space that opens only into a bedroom or bathroom, and such room or space is used for no other purpose and is provided with a solid weather-stripped door equipped with an approved self-closing device. All combustion air shall be taken directly from the outdoors in accordance with Section 304.6.

303.4 Protection from vehicle impact damage. Appliances shall not be installed in a location subject to vehicle impact damage except where protected by an approved means.

303.5 Indoor locations. Furnaces and boilers installed in closets and alcoves shall be listed for such installation.

303.6 Outdoor locations. Equipment installed in outdoor locations shall be either listed for outdoor installation or provided with protection from outdoor environmental factors that influence the operability, durability and safety of the equipment.

303.7 Pit locations. Appliances installed in pits or excavations shall not come in direct contact with the surrounding soil. The sides of the pit or excavation shall be held back a minimum of 12 inches (305 mm) from the appliance. Where the depth exceeds 12 inches (305 mm) below adjoining grade, the walls of the pit or excavation shall be lined with concrete or masonry, such concrete or masonry shall extend a minimum of 4 inches (102 mm) above adjoining grade and shall have sufficient lateral load-bearing capacity to resist collapse. The appliance shall be protected from flooding in an approved manner.

SECTION 304 (IFGS)
COMBUSTION, VENTILATION AND DILUTION AIR

304.1 General. Air for combustion, ventilation and dilution of flue gases for appliances installed in buildings shall be provided by application of one of the methods prescribed in Sections 304.5 through 304.9. Where the requirements of Section 304.5 are not met, outdoor air shall be introduced in accordance with one of the methods prescribed in Sections 304.6 through 304.9. Direct-vent appliances, gas appliances of other than natural draft design and vented gas appliances other than Category I shall be provided with combustion, ventilation and dilution air in accordance with the appliance manufacturer's instructions.

Exception: Type 1 clothes dryers that are provided with makeup air in accordance with Section 614.5.

304.2 Appliance location. Appliances shall be located so as not to interfere with proper circulation of combustion, ventilation and dilution air.

304.3 Draft hood/regulator location. Where used, a draft hood or a barometric draft regulator shall be installed in the same room or enclosure as the appliance served so as to prevent any difference in pressure between the hood or regulator and the combustion air supply.

304.4 Makeup air provisions. Makeup air requirements for the operation of exhaust fans, kitchen ventilation systems, clothes dryers and fireplaces shall be considered in determining the adequacy of a space to provide combustion air requirements.

304.5 Indoor combustion air. The required volume of indoor air shall be determined in accordance with Section 304.5.1 or 304.5.2, except that where the air infiltration rate is known to be less than 0.40 air changes per hour (ACH), Section 304.5.2 shall be used. The total required volume shall be the sum of the required volume calculated for all appliances located within the space. Rooms communicating directly with the space in which the appliances are installed through openings not furnished with doors, and through combustion air openings sized and located in accordance with Section 304.5.3, are considered to be part of the required volume.

304.5.1 Standard method. The minimum required volume shall be 50 cubic feet per 1,000 Btu/h (4.8 m³/kW) of the appliance input rating.

304.5.2 Known air-infiltration-rate method. Where the air infiltration rate of a structure is known, the minimum required volume shall be determined as follows:

For appliances other than fan-assisted, calculate volume using Equation 3-1.

$$\text{Required Volume}_{other} \geq \frac{21\,ft^3}{ACH} \left(\frac{I_{other}}{1,000\,Btu\,/\,hr} \right)$$

(Equation 3-1)

For fan-assisted appliances, calculate volume using Equation 3-2.

$$\text{Required Volume}_{fan} \geq \frac{15\,ft^3}{ACH} \left(\frac{I_{fan}}{1,000\,Btu\,/\,hr} \right)$$

(Equation 3-2)

where:

I_{other} = All appliances other than fan assisted (input in Btu/h).

I_{fan} = Fan-assisted appliance (input in Btu/h).

ACH = Air change per hour (percent of volume of space exchanged per hour, expressed as a decimal).

For purposes of this calculation, an infiltration rate greater than 0.60 ACH shall not be used in Equations 3-1 and 3-2.

304.5.3 Indoor opening size and location. Openings used to connect indoor spaces shall be sized and located in accordance with Sections 304.5.3.1 and 304.5.3.2 (see Figure 304.5.3).

304.5.3.1 Combining spaces on the same story. Each opening shall have a minimum free area of 1 square inch per 1,000 Btu/h (2,200 mm²/kW) of the total input rating of all appliances in the space, but not less than 100 square inches (0.06 m²). One opening shall commence within 12 inches (305 mm) of the top and one opening shall commence within 12 inches (305 mm) of the bottom of

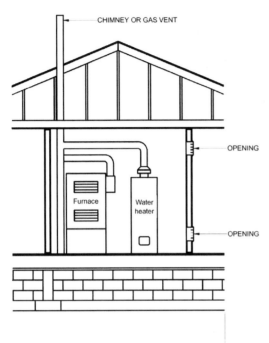

FIGURE 304.5.3
ALL AIR FROM INSIDE THE BUILDING
(see Section 304.5.3)

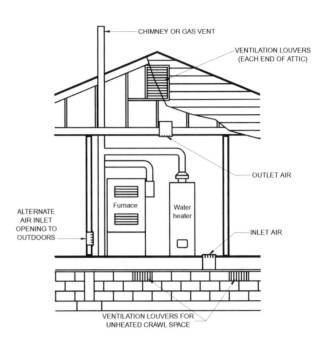

FIGURE 304.6.1(1)
ALL AIR FROM OUTDOORS—INLET AIR FROM VENTILATED
CRAWL SPACE AND OUTLET AIR TO VENTILATED ATTIC
(see Section 304.6.1)

the enclosure. The minimum dimension of air openings shall be not less than 3 inches (76 mm).

304.5.3.2 Combining spaces in different stories. The volumes of spaces in different stories shall be considered as communicating spaces where such spaces are connected by one or more openings in doors or floors having a total minimum free area of 2 square inches per 1,000 Btu/h (4402 mm²/kW) of total input rating of all appliances.

304.6 Outdoor combustion air. Outdoor combustion air shall be provided through opening(s) to the outdoors in accordance with Section 304.6.1 or 304.6.2. The minimum dimension of air openings shall be not less than 3 inches (76 mm).

304.6.1 Two-permanent-openings method. Two permanent openings, one commencing within 12 inches (305 mm) of the top and one commencing within 12 inches (305 mm) of the bottom of the enclosure, shall be provided. The openings shall communicate directly, or by ducts, with the outdoors or spaces that freely communicate with the outdoors.

Where directly communicating with the outdoors, or where communicating with the outdoors through vertical ducts, each opening shall have a minimum free area of 1 square inch per 4,000 Btu/h (550 mm²/kW) of total input rating of all appliances in the enclosure [see Figures 304.6.1(1) and 304.6.1(2)].

Where communicating with the outdoors through horizontal ducts, each opening shall have a minimum free area of not less than 1 square inch per 2,000 Btu/h (1,100 mm²/kW) of total input rating of all appliances in the enclosure [see Figure 304.6.1(3)].

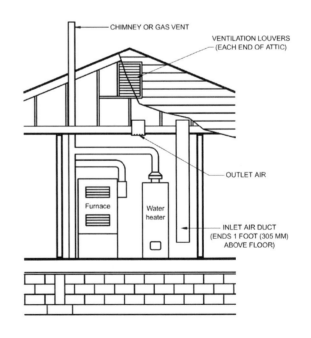

For SI: 1 foot = 304.8 mm.

FIGURE 304.6.1(2)
ALL AIR FROM OUTDOORS THROUGH VENTILATED ATTIC
(see Section 304.6.1)

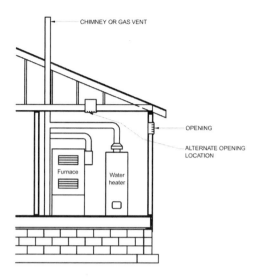

FIGURE 304.6.2
SINGLE COMBUSTION AIR OPENING,
ALL AIR FROM THE OUTDOORS
(see Section 304.6.2)

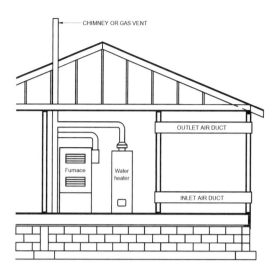

FIGURE 304.6.1(3)
ALL AIR FROM OUTDOORS
(see Section 304.6.1)

304.6.2 One-permanent-opening method. One permanent opening, commencing within 12 inches (305 mm) of the top of the enclosure, shall be provided. The appliance shall have clearances of at least 1 inch (25 mm) from the sides and back and 6 inches (152 mm) from the front of the appliance. The opening shall directly communicate with the outdoors or through a vertical or horizontal duct to the outdoors, or spaces that freely communicate with the outdoors (see Figure 304.6.2) and shall have a minimum free area of 1 square inch per 3,000 Btu/h (734 mm²/kW) of the total input rating of all appliances located in the enclosure and not less than the sum of the areas of all vent connectors in the space.

304.7 Combination indoor and outdoor combustion air. The use of a combination of indoor and outdoor combustion air shall be in accordance with Sections 304.7.1 through 304.7.3.

304.7.1 Indoor openings. Where used, openings connecting the interior spaces shall comply with Section 304.5.3.

304.7.2 Outdoor opening location. Outdoor opening(s) shall be located in accordance with Section 304.6.

304.7.3 Outdoor opening(s) size. The outdoor opening(s) size shall be calculated in accordance with the following:

1. The ratio of interior spaces shall be the available volume of all communicating spaces divided by the required volume.

2. The outdoor size reduction factor shall be one minus the ratio of interior spaces.

3. The minimum size of outdoor opening(s) shall be the full size of outdoor opening(s) calculated in accordance with Section 304.6, multiplied by the reduction factor. The minimum dimension of air openings shall be not less than 3 inches (76 mm).

304.8 Engineered installations. Engineered combustion air installations shall provide an adequate supply of combustion, ventilation and dilution air and shall be approved.

304.9 Mechanical combustion air supply. Where all combustion air is provided by a mechanical air supply system, the combustion air shall be supplied from the outdoors at a rate not less than 0.35 cubic feet per minute per 1,000 Btu/h (0.034 m³/min per kW) of total input rating of all appliances located within the space.

304.9.1 Makeup air. Where exhaust fans are installed, makeup air shall be provided to replace the exhausted air.

304.9.2 Appliance interlock. Each of the appliances served shall be interlocked with the mechanical air supply system to prevent main burner operation when the mechanical air supply system is not in operation.

304.9.3 Combined combustion air and ventilation air system. Where combustion air is provided by the building's mechanical ventilation system, the system shall provide the specified combustion air rate in addition to the required ventilation air.

304.10 Louvers and grilles. The required size of openings for combustion, ventilation and dilution air shall be based on the net free area of each opening. Where the free area through a design of louver, grille or screen is known, it shall be used in calculating the size opening required to provide the free area specified. Where the design and free area of louvers and grilles are not known, it shall be assumed that wood louvers will have 25-percent free area and metal louvers and grilles will have 75-percent free area. Screens shall have a mesh size not smaller than ¹/₄ inch (6.4 mm). Nonmotorized louvers and grilles shall be fixed in the open position. Motorized louvers shall be interlocked with the appliance so that they are proven to be in the full open position prior to main burner ignition and during main burner operation. Means shall be provided to prevent the main burner from igniting if the louvers fail to open during burner start-up and to shut down the main burner if the louvers close during operation.

304.11 Combustion air ducts. Combustion air ducts shall comply with all of the following:

1. Ducts shall be constructed of galvanized steel complying with Chapter 6 of the *International Mechanical Code* or of a material having equivalent corrosion resistance, strength and rigidity.

   Exception: Within dwellings units, unobstructed stud and joist spaces shall not be prohibited from conveying combustion air, provided that not more than one required fireblock is removed.

2. Ducts shall terminate in an unobstructed space allowing free movement of combustion air to the appliances.

3. Ducts shall serve a single enclosure.

4. Ducts shall not serve both upper and lower combustion air openings where both such openings are used. The separation between ducts serving upper and lower combustion air openings shall be maintained to the source of combustion air.

5. Ducts shall not be screened where terminating in an attic space.

6. Horizontal upper combustion air ducts shall not slope downward toward the source of combustion air.

7. The remaining space surrounding a chimney liner, gas vent, special gas vent or plastic piping installed within a masonry, metal or factory-built chimney shall not be used to supply combustion air.

   Exception: Direct-vent gas-fired appliances designed for installation in a solid fuel-burning fireplace where installed in accordance with the manufacturer's instructions.

8. Combustion air intake openings located on the exterior of a building shall have the lowest side of such openings located not less than 12 inches (305 mm) vertically from the adjoining grade level.

304.12 Protection from fumes and gases. Where corrosive or flammable process fumes or gases, other than products of combustion, are present, means for the disposal of such fumes or gases shall be provided. Such fumes or gases include carbon monoxide, hydrogen sulfide, ammonia, chlorine and halogenated hydrocarbons.

In barbershops, beauty shops and other facilities where chemicals that generate corrosive or flammable products, such as aerosol sprays, are routinely used, nondirect vent-type appliances shall be located in a mechanical room separated or partitioned off from other areas with provisions for combustion air and dilution air from the outdoors. Direct-vent appliances shall be installed in accordance with the appliance manufacturer's installation instructions.

SECTION 305 (IFGC)
INSTALLATION

305.1 General. Equipment and appliances shall be installed as required by the terms of their approval, in accordance with the conditions of listing, the manufacturer's instructions and this code. Manufacturers' installation instructions shall be available on the job site at the time of inspection. Where a code provision is less restrictive than the conditions of the listing of the equipment or appliance or the manufacturer's installation instructions, the conditions of the listing and the manufacturer's installation instructions shall apply.

Unlisted appliances approved in accordance with Section 301.3 shall be limited to uses recommended by the manufacturer and shall be installed in accordance with the manufacturer's instructions, the provisions of this code and the requirements determined by the code official.

305.2 Hazardous area. Equipment and appliances having an ignition source shall not be installed in Group H occupancies or control areas where open use, handling or dispensing of combustible, flammable or explosive materials occurs.

305.3 Elevation of ignition source. Equipment and appliances having an ignition source shall be elevated such that the source of ignition is not less than 18 inches (457 mm) above the floor in hazardous locations and public garages, private garages, repair garages, motor fuel-dispensing facilities and parking garages. For the purpose of this section, rooms or spaces that are not part of the living space of a dwelling unit and that communicate directly with a private garage through openings shall be considered to be part of the private garage.

Exception: Elevation of the ignition source is not required for appliances that are listed as flammable vapor ignition resistant.

305.3.1 Parking garages. Connection of a parking garage with any room in which there is a fuel-fired appliance shall be by means of a vestibule providing a two-doorway separation, except that a single door is permitted where the sources of ignition in the appliance are elevated in accordance with Section 305.3.

Exception: This section shall not apply to appliance installations complying with Section 305.4.

305.4 Public garages. Appliances located in public garages, motor fuel-dispensing facilities, repair garages or other areas frequented by motor vehicles shall be installed a minimum of 8 feet (2438 mm) above the floor. Where motor vehicles exceed 6 feet (1829 mm) in height and are capable of passing under an appliance, appliances shall be installed a minimum of 2 feet (610 mm) higher above the floor than the height of the tallest vehicle.

Exception: The requirements of this section shall not apply where the appliances are protected from motor vehicle impact and installed in accordance with Section 305.3 and NFPA 30A.

305.5 Private garages. Appliances located in private garages shall be installed with a minimum clearance of 6 feet (1829 mm) above the floor.

> **Exception:** The requirements of this section shall not apply where the appliances are protected from motor vehicle impact and installed in accordance with Section 305.3.

305.6 Construction and protection. Boiler rooms and furnace rooms shall be protected as required by the *International Building Code*.

305.7 Clearances from grade. Equipment and appliances installed at grade level shall be supported on a level concrete slab or other approved material extending above adjoining grade or shall be suspended a minimum of 6 inches (152 mm) above adjoining grade.

305.8 Clearances to combustible construction. Heat-producing equipment and appliances shall be installed to maintain the required clearances to combustible construction as specified in the listing and manufacturer's instructions. Such clearances shall be reduced only in accordance with Section 308. Clearances to combustibles shall include such considerations as door swing, drawer pull, overhead projections or shelving and window swing. Devices, such as door stops or limits and closers, shall not be used to provide the required clearances.

SECTION 306 (IFGC)
ACCESS AND SERVICE SPACE

[M] 306.1 Clearances for maintenance and replacement. Clearances around appliances to elements of permanent construction, including other installed appliances, shall be sufficient to allow inspection, service, repair or replacement without removing such elements of permanent construction or disabling the function of a required fire-resistance-rated assembly.

[M] 306.2 Appliances in rooms. Rooms containing appliances requiring access shall be provided with a door and an unobstructed passageway measuring not less than 36 inches (914 mm) wide and 80 inches (2032 mm) high.

> **Exception:** Within a dwelling unit, appliances installed in a compartment, alcove, basement or similar space shall be provided with access by an opening or door and an unobstructed passageway measuring not less than 24 inches (610 mm) wide and large enough to allow removal of the largest appliance in the space, provided that a level service space of not less than 30 inches (762 mm) deep and the height of the appliance, but not less than 30 inches (762 mm), is present at the front or service side of the appliance with the door open.

[M] 306.3 Appliances in attics. Attics containing appliances requiring access shall be provided with an opening and unobstructed passageway large enough to allow removal of the largest component of the appliance. The passageway shall not be less than 30 inches (762 mm) high and 22 inches (559 mm) wide and not more than 20 feet (6096 mm) in length when measured along the centerline of the passageway from the opening to the equipment. The passageway shall have continuous solid flooring not less than 24 inches (610 mm) wide. A level service space not less than 30 inches (762 mm) deep and 30 inches (762

mm) wide shall be present at the front or service side of the equipment. The clear access opening dimensions shall be a minimum of 20 inches by 30 inches (508 mm by 762 mm), where such dimensions are large enough to allow removal of the largest component of the appliance.

> **Exceptions:**
> 1. The passageway and level service space are not required where the appliance is capable of being serviced and removed through the required opening.
> 2. Where the passageway is not less than 6 feet (1829 mm) high for its entire length, the passageway shall be not greater than 50 feet (15 250 mm) in length.

[M] 306.3.1 Electrical requirements. A luminaire controlled by a switch located at the required passageway opening and a receptacle outlet shall be provided at or near the equipment location in accordance with the ICC *Electrical Code*.

[M] 306.4 Appliances under floors. Under-floor spaces containing appliances requiring access shall be provided with an access opening and unobstructed passageway large enough to remove the largest component of the appliance. The passageway shall not be less than 30 inches (762 mm) high and 22 inches (559 mm) wide, nor more than 20 feet (6096 mm) in length when measured along the centerline of the passageway from the opening to the equipment. A level service space not less than 30 inches (762 mm) deep and 30 inches (762 mm) wide shall be present at the front or service side of the appliance. If the depth of the passageway or the service space exceeds 12 inches (305 mm) below the adjoining grade, the walls of the passageway shall be lined with concrete or masonry extending 4 inches (102 mm) above the adjoining grade and having sufficient lateral-bearing capacity to resist collapse. The clear access opening dimensions shall be a minimum of 22 inches by 30 inches (559 mm by 762 mm), where such dimensions are large enough to allow removal of the largest component of the appliance.

> **Exceptions:**
> 1. The passageway is not required where the level service space is present when the access is open and the appliance is capable of being serviced and removed through the required opening.
> 2. Where the passageway is not less than 6 feet high (1829 mm) for its entire length, the passageway shall not be limited in length.

[M] 306.4.1 Electrical requirements. A luminaire controlled by a switch located at the required passageway opening and a receptacle outlet shall be provided at or near the equipment location in accordance with the ICC *Electrical Code*.

[M] 306.5 Appliances on roofs or elevated structures. Where appliances requiring access are installed on roofs or elevated structures at a height exceeding 16 feet (4877 mm), such access shall be provided by a permanent approved means of access, the extent of which shall be from grade or floor level to the appliance's level service space. Such access shall not require climbing over obstructions greater than 30 inches high (762 mm) or walking on roofs having a slope greater than four units vertical in 12 units horizontal (33-percent slope).

Permanent ladders installed to provide the required access shall comply with the following minimum design criteria.

1. The side railing shall extend above the parapet or roof edge not less than 30 inches (762 mm).

2. Ladders shall have a rung spacing not to exceed 14 inches (356 mm) on center.

3. Ladders shall have a toe spacing not less than 6 inches (152 mm) deep.

4. There shall be a minimum of 18 inches (457 mm) between rails.

5. Rungs shall have a minimum diameter of 0.75-inch (19 mm) and shall be capable of withstanding a 300-pound (136.1 kg) load.

6. Ladders over 30 feet (9144 mm) in height shall be provided with offset sections and landings capable of withstanding a load of 100 pounds per square foot (488.2 kg/m^2).

7. Ladders shall be protected against corrosion by approved means.

Catwalks installed to provide the required access shall be not less than 24 inches wide (610 mm) and shall have railings as required for service platforms.

Exception: This section shall not apply to Group R-3 occupancies.

[M] 306.5.1 Sloped roofs. Where appliances are installed on a roof having a slope of three units vertical in 12 units horizontal (25-percent slope) or greater and having an edge more than 30 inches (762 mm) above grade at such edge, a level platform shall be provided on each side of the appliance to which access is required for service, repair or maintenance. The platform shall not be less than 30 inches (762 mm) in any dimension and shall be provided with guards. The guards shall extend not less than 42 inches (1067 mm) above the platform, shall be constructed so as to prevent the passage of a 21-inch-diameter (533 mm) sphere and shall comply with the loading requirements for guards specified in the *International Building Code*.

[M] 306.5.2 Electrical requirements. A receptacle outlet shall be provided at or near the equipment location in accordance with the ICC *Electrical Code*.

[M] 306.6 Guards. Guards shall be provided where appliances or other components that require service and roof hatch openings are located within 10 feet (3048 mm) of a roof edge or open side of a walking surface and such edge or open side is located more than 30 inches (762 mm) above the floor, roof or grade below. The guard shall extend not less than 30 inches (762 mm) beyond each end of such appliances, components and roof hatch openings and the top of the guard shall be located not less than 42 inches (1067 mm) above the elevated surface adjacent to the guard. The guard shall be constructed so as to prevent the passage of a 21-inch-diameter (533 mm) sphere and shall comply with the loading requirements for guards specified in the *International Building Code*.

SECTION 307 (IFGC)
CONDENSATE DISPOSAL

307.1 Evaporators and cooling coils. Condensate drainage systems shall be provided for equipment and appliances containing evaporators and cooling coils in accordance with the *International Mechanical Code*.

307.2 Fuel-burning appliances. Liquid combustion by-products of condensing appliances shall be collected and discharged to an approved plumbing fixture or disposal area in accordance with the manufacturer's installation instructions. Condensate piping shall be of approved corrosion-resistant material and shall not be smaller than the drain connection on the appliance. Such piping shall maintain a minimum slope in the direction of discharge of not less than one-eighth unit vertical in 12 units horizontal (1-percent slope).

[M] 307.3 Drain pipe materials and sizes. Components of the condensate disposal system shall be cast iron, galvanized steel, copper, polybutylene, polyethylene, ABS, CPVC or PVC pipe or tubing. All components shall be selected for the pressure and temperature rating of the installation. Condensate waste and drain line size shall be not less than $^3/_4$-inch internal diameter (19 mm) and shall not decrease in size from the drain connection to the place of condensate disposal. Where the drain pipes from more than one unit are manifolded together for condensate drainage, the pipe or tubing shall be sized in accordance with an approved method. All horizontal sections of drain piping shall be installed in uniform alignment at a uniform slope.

307.4 Traps. Condensate drains shall be trapped as required by the equipment or appliance manufacturer.

307.5 Auxiliary drain pan. Category IV condensing appliances shall be provided with an auxiliary drain pan where damage to any building component will occur as a result of stoppage in the condensate drainage system. Such pan shall be installed in accordance with the applicable provisions of Section 307 of the *International Mechanical Code*.

Exception: An auxiliary drain pan shall not be required for appliances that automatically shut down operation in the event of a stoppage in the condensate drainage system.

SECTION 308 (IFGS)
CLEARANCE REDUCTION

308.1 Scope. This section shall govern the reduction in required clearances to combustible materials and combustible assemblies for chimneys, vents, appliances, devices and equipment. Clearance requirements for air-conditioning equipment and central heating boilers and furnaces shall comply with Sections 308.3 and 308.4.

308.2 Reduction table. The allowable clearance reduction shall be based on one of the methods specified in Table 308.2 or shall utilize an assembly listed for such application. Where required clearances are not listed in Table 308.2, the reduced clearances shall be determined by linear interpolation between the distances listed in the table. Reduced clearances shall not be derived by extrapolation below the range of the table. The reduction of the required clearances to combustibles for listed and labeled appliances and equipment shall be in accordance with the requirements of this section except that such clearances shall not be reduced where reduction is specifically prohibited by the terms of the appliance or equipment listing [see Figures 308.2(1) through 308.2(3)].

308.3 Clearances for indoor air-conditioning appliances.
Clearance requirements for indoor air-conditioning appliances shall comply with Sections 308.3.1 through 308.3.5.

308.3.1 Appliances installed in rooms that are large in comparison with the size of the appliance. Air-conditioning appliances installed in rooms that are large in comparison with the size of the appliance shall be installed with clearances in accordance with the manufacturer's instructions.

TABLE 308.2[a through k]
REDUCTION OF CLEARANCES WITH SPECIFIED FORMS OF PROTECTION

TYPE OF PROTECTION APPLIED TO AND COVERING ALL SURFACES OF COMBUSTIBLE MATERIAL WITHIN THE DISTANCE SPECIFIED AS THE REQUIRED CLEARANCE WITH NO PROTECTION [see Figures 308.2(1), 308.2(2), and 308.2(3)]	WHERE THE REQUIRED CLEARANCE WITH NO PROTECTION FROM APPLIANCE, VENT CONNECTOR, OR SINGLE-WALL METAL PIPE IS: (inches)									
	36		18		12		9		6	
	Allowable clearances with specified protection (inches)									
	Use Column 1 for clearances above appliance or horizontal connector. Use Column 2 for clearances from appliance, vertical connector, and single-wall metal pipe.									
	Above Col. 1	Sides and rear Col. 2	Above Col. 1	Sides and rear Col. 2	Above Col. 1	Sides and rear Col. 2	Above Col. 1	Sides and rear Col. 2	Above Col. 1	Sides and rear Col. 2
1. 3½-inch-thick masonry wall without ventilated airspace	—	24	—	12	—	9	—	6	—	5
2. ½-inch insulation board over 1-inch glass fiber or mineral wool batts	24	18	12	9	9	6	6	5	4	3
3. 0.024-inch (nominal 24 gage) sheet metal over 1-inch glass fiber or mineral wool batts reinforced with wire on rear face with ventilated airspace	18	12	9	6	6	4	5	3	3	3
4. 3½-inch-thick masonry wall with ventilated airspace	—	12	—	6	—	6	—	6	—	6
5. 0.024-inch (nominal 24 gage) sheet metal with ventilated airspace	18	12	9	6	6	4	5	3	3	2
6. ½-inch-thick insulation board with ventilated airspace	18	12	9	6	6	4	5	3	3	3
7. 0.024-inch (nominal 24 gage) sheet metal with ventilated airspace over 0.024-inch (nominal 24 gage) sheet metal with ventilated airspace	18	12	9	6	6	4	5	3	3	3
8. 1-inch glass fiber or mineral wool batts sandwiched between two sheets 0.024-inch (nominal 24 gage) sheet metal with ventilated airspace	18	12	9	6	6	4	5	3	3	3

For SI: 1 inch = 25.4 mm, °C = [(°F - 32)/1.8], 1 pound per cubic foot = 16.02 kg/m^3, 1 Btu per inch per square foot per hour per °F = 0.144 W/m^2 × K.

a. Reduction of clearances from combustible materials shall not interfere with combustion air, draft hood clearance and relief, and accessibility of servicing.

b. All clearances shall be measured from the outer surface of the combustible material to the nearest point on the surface of the appliance, disregarding any intervening protection applied to the combustible material.

c. Spacers and ties shall be of noncombustible material. No spacer or tie shall be used directly opposite an appliance or connector.

d. For all clearance reduction systems using a ventilated airspace, adequate provision for air circulation shall be provided as described [see Figures 308.2(2) and 308.2(3)].

e. There shall be at least 1 inch between clearance reduction systems and combustible walls and ceilings for reduction systems using ventilated airspace.

f. Where a wall protector is mounted on a single flat wall away from corners, it shall have a minimum 1-inch air gap. To provide air circulation, the bottom and top edges, or only the side and top edges, or all edges shall be left open.

g. Mineral wool batts (blanket or board) shall have a minimum density of 8 pounds per cubic foot and a minimum melting point of 1500°F.

h. Insulation material used as part of a clearance reduction system shall have a thermal conductivity of 1.0 Btu per inch per square foot per hour per °F or less.

i. There shall be at least 1 inch between the appliance and the protector. In no case shall the clearance between the appliance and the combustible surface be reduced below that allowed in this table.

j. All clearances and thicknesses are minimum; larger clearances and thicknesses are acceptable.

k. Listed single-wall connectors shall be installed in accordance with the manufacturer's installation instructions.

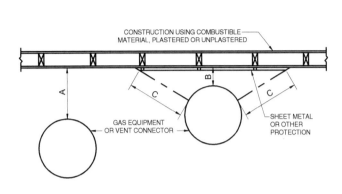

"A" equals the reduced clearance with no protection.

"B" equals the reduced clearance permitted in accordance with Table 308.2. The protection applied to the construction using combustible material shall extend far enough in each direction to make "C" equal to "A."

**FIGURE 308.2(1)
EXTENT OF PROTECTION NECESSARY TO
REDUCE CLEARANCES FROM APPLIANCE OR
VENT CONNECTIONS**

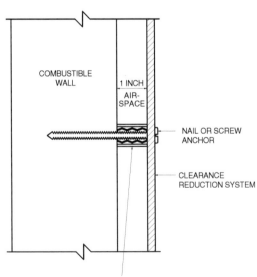

For SI: 1 inch = 25.4 mm.

**FIGURE 308.2(3)
MASONRY CLEARANCE REDUCTION SYSTEM**

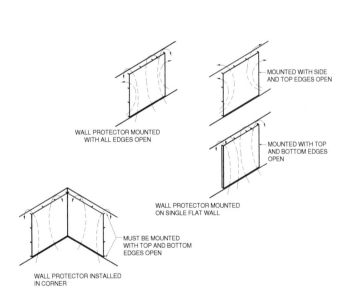

For SI: 1 inch = 25.4 mm.

**FIGURE 308.2(2)
WALL PROTECTOR CLEARANCE REDUCTION SYSTEM**

308.3.2 Appliances installed in rooms that are not large in comparison with the size of the appliance. Air-conditioning appliances installed in rooms that are not large in comparison with the size of the appliance, such as alcoves and closets, shall be listed for such installations and installed in accordance with the manufacturer's instructions. Listed clearances shall not be reduced by the protection methods described in Table 308.2, regardless of whether the enclosure is of combustible or noncombustible material.

308.3.3 Clearance reduction. Air-conditioning appliances installed in rooms that are large in comparison with the size of the appliance shall be permitted to be installed with reduced clearances to combustible material, provided the combustible material or appliance is protected as described in Table 308.2.

308.3.4 Plenum clearances. Where the furnace plenum is adjacent to plaster on metal lath or noncombustible material attached to combustible material, the clearance shall be measured to the surface of the plaster or other noncombustible finish where the clearance specified is 2 inches (51 mm) or less.

308.3.5 Clearance from supply ducts. Air-conditioning appliances shall have the clearance from supply ducts within 3 feet (914 mm) of the furnace plenum be not less than that specified from the furnace plenum. Clearance is not necessary beyond this distance.

308.4 Central-heating boilers and furnaces. Clearance requirements for central-heating boilers and furnaces shall comply with Sections 308.4.1 through 308.4.6. The clearance to these appliances shall not interfere with combustion air; draft hood clearance and relief; and accessibility for servicing.

308.4.1 Appliances installed in rooms that are large in comparison with the size of the appliance. Central-heating furnaces and low-pressure boilers installed in rooms large in comparison with the size of the appliance shall be installed with clearances in accordance with the manufacturer's instructions.

308.4.2 Appliances installed in rooms that are not large in comparison with the size of the appliance. Central-heating furnaces and low-pressure boilers installed in rooms that are not large in comparison with the size of the appliance, such as alcoves and closets, shall be listed for such installations. Listed clearances shall not be reduced by the protection methods described in Table 308.2 and illustrated in Figures 308.2(1) through 308.2(3), regardless of whether the enclosure is of combustible or noncombustible material.

308.4.3 Clearance reduction. Central-heating furnaces and low-pressure boilers installed in rooms that are large in comparison with the size of the appliance shall be permitted to be installed with reduced clearances to combustible material provided the combustible material or appliance is protected as described in Table 308.2.

308.4.4 Clearance for servicing appliances. Front clearance shall be sufficient for servicing the burner and the furnace or boiler.

308.4.5 Plenum clearances. Where the furnace plenum is adjacent to plaster on metal lath or noncombustible material attached to combustible material, the clearance shall be measured to the surface of the plaster or other noncombustible finish where the clearance specified is 2 inches (51 mm) or less.

308.4.6 Clearance from supply ducts. Central-heating furnaces shall have the clearance from supply ducts within 3 feet (914 mm) of the furnace plenum be not less than that specified from the furnace plenum. No clearance is necessary beyond this distance.

SECTION 309 (IFGC)
ELECTRICAL

309.1 Grounding. Gas piping shall not be used as a grounding electrode.

309.2 Connections. Electrical connections between equipment and the building wiring, including the grounding of the equipment, shall conform to the ICC *Electrical Code.*

SECTION 310 (IFGS)
ELECTRICAL BONDING

310.1 Gas pipe bonding. Each above-ground portion of a gas piping system that is likely to become energized shall be electrically continuous and bonded to an effective ground-fault current path. Gas piping shall be considered to be bonded where it is connected to appliances that are connected to the equipment grounding conductor of the circuit supplying that appliance.

CHAPTER 4

GAS PIPING INSTALLATIONS

SECTION 401 (IFGC)
GENERAL

401.1 Scope. This chapter shall govern the design, installation, modification and maintenance of piping systems. The applicability of this code to piping systems extends from the point of delivery to the connections with the equipment and includes the design, materials, components, fabrication, assembly, installation, testing, inspection, operation and maintenance of such piping systems.

401.1.1 Utility piping systems located within buildings. Utility service piping located within buildings shall be installed in accordance with the structural safety and fire protection provisions of the *International Building Code*.

401.2 Liquefied petroleum gas storage. The storage system for liquefied petroleum gas shall be designed and installed in accordance with the *International Fire Code* and NFPA 58.

401.3 Modifications to existing systems. In modifying or adding to existing piping systems, sizes shall be maintained in accordance with this chapter.

401.4 Additional appliances. Where an additional appliance is to be served, the existing piping shall be checked to determine if it has adequate capacity for all appliances served. If inadequate, the existing system shall be enlarged as required or separate piping of adequate capacity shall be provided.

401.5 Identification. For other than steel pipe, exposed piping shall be identified by a yellow label marked "Gas" in black letters. The marking shall be spaced at intervals not exceeding 5 feet (1524 mm). The marking shall not be required on pipe located in the same room as the equipment served.

401.6 Interconnections. Where two or more meters are installed on the same premises but supply separate consumers, the piping systems shall not be interconnected on the outlet side of the meters.

401.7 Piping meter identification. Piping from multiple meter installations shall be marked with an approved permanent identification by the installer so that the piping system supplied by each meter is readily identifiable.

401.8 Minimum sizes. All pipe utilized for the installation, extension and alteration of any piping system shall be sized to supply the full number of outlets for the intended purpose and shall be sized in accordance with Section 402.

SECTION 402 (IFGS)
PIPE SIZING

402.1 General considerations. Piping systems shall be of such size and so installed as to provide a supply of gas sufficient to meet the maximum demand without undue loss of pressure between the point of delivery and the appliance.

402.2 Maximum gas demand. The volume of gas to be provided, in cubic feet per hour, shall be determined directly from the manufacturer's input ratings of the appliances served. Where an input rating is not indicated, the gas supplier, appliance manufacturer or a qualified agency shall be contacted, or the rating from Table 402.2 shall be used for estimating the volume of gas to be supplied.

The total connected hourly load shall be used as the basis for pipe sizing, assuming that all appliances could be operating at full capacity simultaneously. Where a diversity of load can be established, pipe sizing shall be permitted to be based on such loads.

TABLE 402.2
APPROXIMATE GAS INPUT FOR TYPICAL APPLIANCES

APPLIANCE	INPUT BTU/H (Approx.)
Space Heating Units	
Hydronic boiler	
Single family	100,000
Multifamily, per unit	60,000
Warm-air furnace	
Single family	100,000
Multifamily, per unit	60,000
Space and Water Heating Units	
Hydronic boiler	
Single family	120,000
Multifamily, per unit	75,000
Water Heating Appliances	
Water heater, automatic instantaneous	
Capacity at 2 gal./minute	142,800
Capacity at 4 gal./minute	285,000
Capacity at 6 gal./minute	428,400
Water heater, automatic storage, 30- to 40-gal. tank	35,000
Water heater, automatic storage, 50-gal. tank	50,000
Water heater, domestic, circulating or side-arm	35,000
Cooking Appliances	
Built-in oven or broiler unit, domestic	25,000
Built-in top unit, domestic	40,000
Range, free-standing, domestic	65,000
Other Appliances	
Barbecue	40,000
Clothes dryer, Type 1 (domestic)	35,000
Gas fireplace, direct-vent	40,000
Gas light	2,500
Gas log	80,000
Refrigerator	3,000

For SI: 1 British thermal unit per hour = 0.293 W, 1 gallon = 3.785 L, 1 gallon per minute = 3.785 L/m.

402.3 Sizing. Gas piping shall be sized in accordance with one of the following:

1. Pipe sizing tables or sizing equations in accordance with Section 402.4.

2. The sizing tables included in a listed piping system's manufacturer's installation instructions.

3. Other approved engineering methods.

402.4 Sizing tables and equations. Where Tables 402.4(1) through 402.4(35) are used to size piping or tubing, the pipe length shall be determined in accordance with Section 402.4.1, 402.4.2 or 402.4.3.

Where Equations 4-1 and 4-2 are used to size piping or tubing, the pipe or tubing shall have smooth inside walls and the pipe length shall be determined in accordance with Section 402.4.1, 402.4.2 or 402.4.3.

1. Low-pressure gas equation [Less than 1.5 pounds per square inch (psi) (10.3 kPa)]:

$$D = \frac{Q^{0.381}}{19.17 \left(\dfrac{\Delta H}{C_r \times L} \right)^{0.206}}$$ **(Equation 4-1)**

2. High-pressure gas equation [1.5 psi (10.3 kPa) and above]:

$$D = \frac{Q^{0.381}}{18.93 \left[\dfrac{\left(P_1^{\,2} - P_2^{\,2} \right) \times Y}{C_r \times L} \right]^{0.206}}$$ **(Equation 4-2)**

where:

D = Inside diameter of pipe, inches (mm).

Q = Input rate appliance(s), cubic feet per hour at 60°F (16°C) and 30-inch mercury column

P_1 = Upstream pressure, psia (P_1 + 14.7)

P_2 = Downstream pressure, psia (P_2 + 14.7)

L = Equivalent length of pipe, feet

ΔH = Pressure drop, inch water column (27.7 inch water column = 1 psi)

TABLE 402.4
C_r AND Y VALUES FOR NATURAL GAS AND
UNDILUTED PROPANE AT STANDARD CONDITIONS

GAS	EQUATION FACTORS	
	C_r	Y
Natural gas	0.6094	0.9992
Undiluted propane	1.2462	0.9910

For SI: 1 cubic foot = 0.028 m³, 1 foot = 305 mm, 1-inch water column = 0.249 kPa, 1 pound per square inch = 6.895 kPa, 1 British thermal unit per hour = 0.293 W.

402.4.1 Longest length method. The pipe size of each section of gas piping shall be determined using the longest length of piping from the point of delivery to the most remote outlet and the load of the section.

402.4.2 Branch length method. Pipe shall be sized as follows:

1. Pipe size of each section of the longest pipe run from the point of delivery to the most remote outlet shall be determined using the longest run of piping and the load of the section.

2. The pipe size of each section of branch piping not previously sized shall be determined using the length of piping from the point of delivery to the most remote outlet in each branch and the load of the section.

402.4.3 Hybrid pressure. The pipe size for each section of higher pressure gas piping shall be determined using the longest length of piping from the point of delivery to the most remote line pressure regulator. The pipe size from the line pressure regulator to each outlet shall be determined using the length of piping from the regulator to the most remote outlet served by the regulator.

402.5 Allowable pressure drop. The design pressure loss in any piping system under maximum probable flow conditions, from the point of delivery to the inlet connection of the appliance, shall be such that the supply pressure at the appliance is greater than the minimum pressure required for proper appliance operation.

402.6 Maximum design operating pressure. The maximum design operating pressure for piping systems located inside buildings shall not exceed 5 pounds per square inch gauge (psig) (34 kPa gauge) except where one or more of the following conditions are met:

1. The piping system is welded.

2. The piping is located in a ventilated chase or otherwise enclosed for protection against accidental gas accumulation.

3. The piping is located inside buildings or separate areas of buildings used exclusively for:

    3.1. Industrial processing or heating;

    3.2. Research;

    3.3. Warehousing; or

    3.4. Boiler or mechanical rooms.

4. The piping is a temporary installation for buildings under construction.

402.6.1 Liquefied petroleum gas systems. The operating pressure for undiluted LP-gas systems shall not exceed 20 psig (140 kPa gauge). Buildings having systems designed to operate below -5°F (-21°C) or with butane or a propane-butane mix shall be designed to either accommodate liquid LP-gas or prevent LP-gas vapor from condensing into a liquid.

Exception: Buildings or separate areas of buildings constructed in accordance with Chapter 10 of NFPA 58 and used exclusively to house industrial processes, research and experimental laboratories, or equipment or processing having similar hazards.

TABLE 402.4(1)
SCHEDULE 40 METALLIC PIPE

Gas	Natural
Inlet Pressure	Less than 2 psi
Pressure Drop	0.3 in. w.c.
Specific Gravity	0.60

PIPE SIZE (inch)														
Nominal	$^1/_2$	$^3/_4$	1	$1^1/_4$	$1^1/_2$	2	$2^1/_2$	3	4	5	6	8	10	12
Actual ID	0.622	0.824	1.049	1.380	1.610	2.067	2.469	3.068	4.026	5.047	6.065	7.981	10.020	11.938
Length (ft)	Capacity in Cubic Feet of Gas Per Hour													
10	131	273	514	1,060	1,580	3,050	4,860	8,580	17,500	31,700	51,300	105,000	191,000	303,000
20	90	188	353	726	1,090	2,090	3,340	5,900	12,000	21,800	35,300	72,400	132,000	208,000
30	72	151	284	583	873	1,680	2,680	4,740	9,660	17,500	28,300	58,200	106,000	167,000
40	62	129	243	499	747	1,440	2,290	4,050	8,270	15,000	24,200	49,800	90,400	143,000
50	55	114	215	442	662	1,280	2,030	3,590	7,330	13,300	21,500	44,100	80,100	127,000
60	50	104	195	400	600	1,160	1,840	3,260	6,640	12,000	19,500	40,000	72,600	115,000
70	46	95	179	368	552	1,060	1,690	3,000	6,110	11,100	17,900	36,800	66,800	106,000
80	42	89	167	343	514	989	1,580	2,790	5,680	10,300	16,700	34,200	62,100	98,400
90	40	83	157	322	482	928	1,480	2,610	5,330	9,650	15,600	32,100	58,300	92,300
100	38	79	148	304	455	877	1,400	2,470	5,040	9,110	14,800	30,300	55,100	87,200
125	33	70	131	269	403	777	1,240	2,190	4,460	8,080	13,100	26,900	48,800	77,300
150	30	63	119	244	366	704	1,120	1,980	4,050	7,320	11,900	24,300	44,200	70,000
175	28	58	109	224	336	648	1,030	1,820	3,720	6,730	10,900	22,400	40,700	64,400
200	26	54	102	209	313	602	960	1,700	3,460	6,260	10,100	20,800	37,900	59,900
250	23	48	90	185	277	534	851	1,500	3,070	5,550	8,990	18,500	33,500	53,100
300	21	43	82	168	251	484	771	1,360	2,780	5,030	8,150	16,700	30,400	48,100
350	19	40	75	154	231	445	709	1,250	2,560	4,630	7,490	15,400	28,000	44,300
400	18	37	70	143	215	414	660	1,170	2,380	4,310	6,970	14,300	26,000	41,200
450	17	35	66	135	202	389	619	1,090	2,230	4,040	6,540	13,400	24,400	38,600
500	16	33	62	127	191	367	585	1,030	2,110	3,820	6,180	12,700	23,100	36,500
550	15	31	59	121	181	349	556	982	2,000	3,620	5,870	12,100	21,900	34,700
600	14	30	56	115	173	333	530	937	1,910	3,460	5,600	11,500	20,900	33,100
650	14	29	54	110	165	318	508	897	1,830	3,310	5,360	11,000	20,000	31,700
700	13	27	52	106	159	306	488	862	1,760	3,180	5,150	10,600	19,200	30,400
750	13	26	50	102	153	295	470	830	1,690	3,060	4,960	10,200	18,500	29,300
800	12	26	48	99	148	285	454	802	1,640	2,960	4,790	9,840	17,900	28,300
850	12	25	46	95	143	275	439	776	1,580	2,860	4,640	9,530	17,300	27,400
900	11	24	45	93	139	267	426	752	1,530	2,780	4,500	9,240	16,800	26,600
950	11	23	44	90	135	259	413	731	1,490	2,700	4,370	8,970	16,300	25,800
1,000	11	23	43	87	131	252	402	711	1,450	2,620	4,250	8,720	15,800	25,100
1,100	10	21	40	83	124	240	382	675	1,380	2,490	4,030	8,290	15,100	23,800
1,200	NA	20	39	79	119	229	364	644	1,310	2,380	3,850	7,910	14,400	22,700
1,300	NA	20	37	76	114	219	349	617	1,260	2,280	3,680	7,570	13,700	21,800
1,400	NA	19	35	73	109	210	335	592	1,210	2,190	3,540	7,270	13,200	20,900
1,500	NA	18	34	70	105	203	323	571	1,160	2,110	3,410	7,010	12,700	20,100
1,600	NA	18	33	68	102	196	312	551	1,120	2,030	3,290	6,770	12,300	19,500
1,700	NA	17	32	66	98	189	302	533	1,090	1,970	3,190	6,550	11,900	18,800
1,800	NA	16	31	64	95	184	293	517	1,050	1,910	3,090	6,350	11,500	18,300
1,900	NA	16	30	62	93	178	284	502	1,020	1,850	3,000	6,170	11,200	17,700
2,000	NA	16	29	60	90	173	276	488	1,000	1,800	2,920	6,000	10,900	17,200

For SI: 1 inch = 25.4 mm, 1 foot = 304.8 mm, 1 pound per square inch = 6.895 kPa, 1-inch water column = 0.2488 kPa,
1 British thermal unit per hour = 0.2931 W, 1 cubic foot per hour = 0.0283 m³/h, 1 degree = 0.01745 rad.

Notes:
1. NA means a flow of less than 10 cfh.
2. All table entries have been rounded to three significant digits.

TABLE 402.4(2)
SCHEDULE 40 METALLIC PIPE

Gas	Natural
Inlet Pressure	Less than 2 psi
Pressure Drop	0.5 in. w.c.
Specific Gravity	0.60

PIPE SIZE (inch)														
Nominal	$^1/_2$	$^3/_4$	1	$1^1/_4$	$1^1/_2$	2	$2^1/_2$	3	4	5	6	8	10	12
Actual ID	0.622	0.824	1.049	1.380	1.610	2.067	2.469	3.068	4.026	5.047	6.065	7.981	10.020	11.938
Length (ft)	Capacity in Cubic Feet of Gas Per Hour													
10	172	360	678	1,390	2,090	4,020	6,400	11,300	23,100	41,800	67,600	139,000	252,000	399,000
20	118	247	466	957	1,430	2,760	4,400	7,780	15,900	28,700	46,500	95,500	173,000	275,000
30	95	199	374	768	1,150	2,220	3,530	6,250	12,700	23,000	37,300	76,700	139,000	220,000
40	81	170	320	657	985	1,900	3,020	5,350	10,900	19,700	31,900	65,600	119,000	189,000
50	72	151	284	583	873	1,680	2,680	4,740	9,660	17,500	28,300	58,200	106,000	167,000
60	65	137	257	528	791	1,520	2,430	4,290	8,760	15,800	25,600	52,700	95,700	152,000
70	60	126	237	486	728	1,400	2,230	3,950	8,050	14,600	23,600	48,500	88,100	139,000
80	56	117	220	452	677	1,300	2,080	3,670	7,490	13,600	22,000	45,100	81,900	130,000
90	52	110	207	424	635	1,220	1,950	3,450	7,030	12,700	20,600	42,300	76,900	122,000
100	50	104	195	400	600	1,160	1,840	3,260	6,640	12,000	19,500	40,000	72,600	115,000
125	44	92	173	355	532	1,020	1,630	2,890	5,890	10,600	17,200	35,400	64,300	102,000
150	40	83	157	322	482	928	1,480	2,610	5,330	9,650	15,600	32,100	58,300	92,300
175	37	77	144	296	443	854	1,360	2,410	4,910	8,880	14,400	29,500	53,600	84,900
200	34	71	134	275	412	794	1,270	2,240	4,560	8,260	13,400	27,500	49,900	79,000
250	30	63	119	244	366	704	1,120	1,980	4,050	7,320	11,900	24,300	44,200	70,000
300	27	57	108	221	331	638	1,020	1,800	3,670	6,630	10,700	22,100	40,100	63,400
350	25	53	99	203	305	587	935	1,650	3,370	6,100	9,880	20,300	36,900	58,400
400	23	49	92	189	283	546	870	1,540	3,140	5,680	9,190	18,900	34,300	54,300
450	22	46	86	177	266	512	816	1,440	2,940	5,330	8,620	17,700	32,200	50,900
500	21	43	82	168	251	484	771	1,360	2,780	5,030	8,150	16,700	30,400	48,100
550	20	41	78	159	239	459	732	1,290	2,640	4,780	7,740	15,900	28,900	45,700
600	19	39	74	152	228	438	699	1,240	2,520	4,560	7,380	15,200	27,500	43,600
650	18	38	71	145	218	420	669	1,180	2,410	4,360	7,070	14,500	26,400	41,800
700	17	36	68	140	209	403	643	1,140	2,320	4,190	6,790	14,000	25,300	40,100
750	17	35	66	135	202	389	619	1,090	2,230	4,040	6,540	13,400	24,400	38,600
800	16	34	63	130	195	375	598	1,060	2,160	3,900	6,320	13,000	23,600	37,300
850	16	33	61	126	189	363	579	1,020	2,090	3,780	6,110	12,600	22,800	36,100
900	15	32	59	122	183	352	561	992	2,020	3,660	5,930	12,200	22,100	35,000
950	15	31	58	118	178	342	545	963	1,960	3,550	5,760	11,800	21,500	34,000
1,000	14	30	56	115	173	333	530	937	1,910	3,460	5,600	11,500	20,900	33,100
1,100	14	28	53	109	164	316	503	890	1,810	3,280	5,320	10,900	19,800	31,400
1,200	13	27	51	104	156	301	480	849	1,730	3,130	5,070	10,400	18,900	30,000
1,300	12	26	49	100	150	289	460	813	1,660	3,000	4,860	9,980	18,100	28,700
1,400	12	25	47	96	144	277	442	781	1,590	2,880	4,670	9,590	17,400	27,600
1,500	11	24	45	93	139	267	426	752	1,530	2,780	4,500	9,240	16,800	26,600
1,600	11	23	44	89	134	258	411	727	1,480	2,680	4,340	8,920	16,200	25,600
1,700	11	22	42	86	130	250	398	703	1,430	2,590	4,200	8,630	15,700	24,800
1,800	10	22	41	84	126	242	386	682	1,390	2,520	4,070	8,370	15,200	24,100
1,900	10	21	40	81	122	235	375	662	1,350	2,440	3,960	8,130	14,800	23,400
2,000	NA	20	39	79	119	229	364	644	1,310	2,380	3,850	7,910	14,400	22,700

For SI: 1 inch = 25.4 mm, 1 foot = 304.8 mm, 1 pound per square inch = 6.895 kPa, 1-inch water column = 0.2488 kPa,
1 British thermal unit per hour = 0.2931 W, 1 cubic foot per hour = 0.0283 m³/h, 1 degree = 0.01745 rad.

Notes:
1. NA means a flow of less than 10 cfh.
2. All table entries have been rounded to three significant digits.

TABLE 402.4(3)
SCHEDULE 40 METALLIC PIPE

	Gas	Natural
	Inlet Pressure	2.0 psi
	Pressure Drop	1.0 psi
	Specific Gravity	0.60

PIPE SIZE (inch)									
Nominal	$^1/_2$	$^3/_4$	1	$1^1/_4$	$1^1/_2$	2	$2^1/_2$	3	4
Actual ID	0.622	0.824	1.049	1.380	1.610	2.067	2.469	3.068	4.026
Length (ft)	Capacity in Cubic Feet of Gas Per Hour								
10	1,510	3,040	5,560	11,400	17,100	32,900	52,500	92,800	189,000
20	1,070	2,150	3,930	8,070	12,100	23,300	37,100	65,600	134,000
30	869	1,760	3,210	6,590	9,880	19,000	30,300	53,600	109,000
40	753	1,520	2,780	5,710	8,550	16,500	26,300	46,400	94,700
50	673	1,360	2,490	5,110	7,650	14,700	23,500	41,500	84,700
60	615	1,240	2,270	4,660	6,980	13,500	21,400	37,900	77,300
70	569	1,150	2,100	4,320	6,470	12,500	19,900	35,100	71,600
80	532	1,080	1,970	4,040	6,050	11,700	18,600	32,800	67,000
90	502	1,010	1,850	3,810	5,700	11,000	17,500	30,900	63,100
100	462	934	1,710	3,510	5,260	10,100	16,100	28,500	58,200
125	414	836	1,530	3,140	4,700	9,060	14,400	25,500	52,100
150	372	751	1,370	2,820	4,220	8,130	13,000	22,900	46,700
175	344	695	1,270	2,601	3,910	7,530	12,000	21,200	43,300
200	318	642	1,170	2,410	3,610	6,960	11,100	19,600	40,000
250	279	583	1,040	2,140	3,210	6,180	9,850	17,400	35,500
300	253	528	945	1,940	2,910	5,600	8,920	15,800	32,200
350	232	486	869	1,790	2,670	5,150	8,210	14,500	29,600
400	216	452	809	1,660	2,490	4,790	7,640	13,500	27,500
450	203	424	759	1,560	2,330	4,500	7,170	12,700	25,800
500	192	401	717	1,470	2,210	4,250	6,770	12,000	24,400
550	182	381	681	1,400	2,090	4,030	6,430	11,400	23,200
600	174	363	650	1,330	2,000	3,850	6,130	10,800	22,100
650	166	348	622	1,280	1,910	3,680	5,870	10,400	21,200
700	160	334	598	1,230	1,840	3,540	5,640	9,970	20,300
750	154	322	576	1,180	1,770	3,410	5,440	9,610	19,600
800	149	311	556	1,140	1,710	3,290	5,250	9,280	18,900
850	144	301	538	1,100	1,650	3,190	5,080	8,980	18,300
900	139	292	522	1,070	1,600	3,090	4,930	8,710	17,800
950	135	283	507	1,040	1,560	3,000	4,780	8,460	17,200
1,000	132	275	493	1,010	1,520	2,920	4,650	8,220	16,800
1,100	125	262	468	960	1,440	2,770	4,420	7,810	15,900
1,200	119	250	446	917	1,370	2,640	4,220	7,450	15,200
1,300	114	239	427	878	1,320	2,530	4,040	7,140	14,600
1,400	110	230	411	843	1,260	2,430	3,880	6,860	14,000
1,500	106	221	396	812	1,220	2,340	3,740	6,600	13,500
1,600	102	214	382	784	1,180	2,260	3,610	6,380	13,000
1,700	99	207	370	759	1,140	2,190	3,490	6,170	12,600
1,800	96	200	358	736	1,100	2,120	3,390	5,980	12,200
1,900	93	195	348	715	1,070	2,060	3,290	5,810	11,900
2,000	91	189	339	695	1,040	2,010	3,200	5,650	11,500

For SI: 1 inch = 25.4 mm, 1 foot = 304.8 mm, 1 pound per square inch = 6.895 kPa, 1-inch water column = 0.2488 kPa,
1 British thermal unit per hour = 0.2931 W, 1 cubic foot per hour = 0.0283 m³/h, 1 degree = 0.01745 rad.
Note: All table entries have been rounded to three significant digits.

TABLE 402.4(4)
SCHEDULE 40 METALLIC PIPE

	Gas	Natural
	Inlet Pressure	3.0 psi
	Pressure Drop	2.0 psi
	Specific Gravity	0.60

	PIPE SIZE (inch)								
Nominal	1/2	3/4	1	1 1/4	1 1/2	2	2 1/2	3	4
Actual ID	0.622	0.824	1.049	1.380	1.610	2.067	2.469	3.068	4.026
Length (ft)	Capacity in Cubic Feet of Gas Per Hour								
10	2,350	4,920	9,270	19,000	28,500	54,900	87,500	155,000	316,000
20	1,620	3,380	6,370	13,100	19,600	37,700	60,100	106,000	217,000
30	1,300	2,720	5,110	10,500	15,700	30,300	48,300	85,400	174,000
40	1,110	2,320	4,380	8,990	13,500	25,900	41,300	73,100	149,000
50	985	2,060	3,880	7,970	11,900	23,000	36,600	64,800	132,000
60	892	1,870	3,520	7,220	10,800	20,800	33,200	58,700	120,000
70	821	1,720	3,230	6,640	9,950	19,200	30,500	54,000	110,000
80	764	1,600	3,010	6,180	9,260	17,800	28,400	50,200	102,000
90	717	1,500	2,820	5,800	8,680	16,700	26,700	47,100	96,100
100	677	1,420	2,670	5,470	8,200	15,800	25,200	44,500	90,800
125	600	1,250	2,360	4,850	7,270	14,000	22,300	39,500	80,500
150	544	1,140	2,140	4,400	6,590	12,700	20,200	35,700	72,900
175	500	1,050	1,970	4,040	6,060	11,700	18,600	32,900	67,100
200	465	973	1,830	3,760	5,640	10,900	17,300	30,600	62,400
250	412	862	1,620	3,330	5,000	9,620	15,300	27,100	55,300
300	374	781	1,470	3,020	4,530	8,720	13,900	24,600	50,100
350	344	719	1,350	2,780	4,170	8,020	12,800	22,600	46,100
400	320	669	1,260	2,590	3,870	7,460	11,900	21,000	42,900
450	300	627	1,180	2,430	3,640	7,000	11,200	19,700	40,200
500	283	593	1,120	2,290	3,430	6,610	10,500	18,600	38,000
550	269	563	1,060	2,180	3,260	6,280	10,000	17,700	36,100
600	257	537	1,010	2,080	3,110	5,990	9,550	16,900	34,400
650	246	514	969	1,990	2,980	5,740	9,150	16,200	33,000
700	236	494	931	1,910	2,860	5,510	8,790	15,500	31,700
750	228	476	897	1,840	2,760	5,310	8,470	15,000	30,500
800	220	460	866	1,780	2,660	5,130	8,180	14,500	29,500
850	213	445	838	1,720	2,580	4,960	7,910	14,000	28,500
900	206	431	812	1,670	2,500	4,810	7,670	13,600	27,700
950	200	419	789	1,620	2,430	4,670	7,450	13,200	26,900
1,000	195	407	767	1,580	2,360	4,550	7,240	12,800	26,100
1,100	185	387	729	1,500	2,240	4,320	6,890	12,200	24,800
1,200	177	369	695	1,430	2,140	4,120	6,570	11,600	23,700
1,300	169	353	666	1,370	2,050	3,940	6,290	11,100	22,700
1,400	162	340	640	1,310	1,970	3,790	6,040	10,700	21,800
1,500	156	327	616	1,270	1,900	3,650	5,820	10,300	21,000
1,600	151	316	595	1,220	1,830	3,530	5,620	10,000	20,300
1,700	146	306	576	1,180	1,770	3,410	5,440	9,610	19,600
1,800	142	296	558	1,150	1,720	3,310	5,270	9,320	19,000
1,900	138	288	542	1,110	1,670	3,210	5,120	9,050	18,400
2,000	134	280	527	1,080	1,620	3,120	4,980	8,800	18,000

For SI: 1 inch = 25.4 mm, 1 foot = 304.8 mm, 1 pound per square inch = 6.895 kPa, 1-inch water column = 0.2488 kPa,
1 British thermal unit per hour = 0.2931 W, 1 cubic foot per hour = 0.0283 m³/h, 1 degree = 0.01745 rad.

Note: All table entries have been rounded to three significant digits.

TABLE 402.4(5)
SCHEDULE 40 METALLIC PIPE

Gas	Natural
Inlet Pressure	5.0 psi
Pressure Drop	3.5 psi
Specific Gravity	0.60

PIPE SIZE (inch)									
Nominal	$^1/_2$	$^3/_4$	1	$1^1/_4$	$1^1/_2$	2	$2^1/_2$	3	4
Actual ID	0.622	0.824	1.049	1.380	1.610	2.067	2.469	3.068	4.026
Length (ft)	Capacity in Cubic Feet of Gas Per Hour								
10	3,190	6,430	11,800	24,200	36,200	69,700	111,000	196,000	401,000
20	2,250	4,550	8,320	17,100	25,600	49,300	78,600	139,000	283,000
30	1,840	3,720	6,790	14,000	20,900	40,300	64,200	113,000	231,000
40	1,590	3,220	5,880	12,100	18,100	34,900	55,600	98,200	200,000
50	1,430	2,880	5,260	10,800	16,200	31,200	49,700	87,900	179,000
60	1,300	2,630	4,800	9,860	14,800	28,500	45,400	80,200	164,000
70	1,200	2,430	4,450	9,130	13,700	26,400	42,000	74,300	151,000
80	1,150	2,330	4,260	8,540	12,800	24,700	39,300	69,500	142,000
90	1,060	2,150	3,920	8,050	12,100	23,200	37,000	65,500	134,000
100	979	1,980	3,620	7,430	11,100	21,400	34,200	60,400	123,000
125	876	1,770	3,240	6,640	9,950	19,200	30,600	54,000	110,000
150	786	1,590	2,910	5,960	8,940	17,200	27,400	48,500	98,900
175	728	1,470	2,690	5,520	8,270	15,900	25,400	44,900	91,600
200	673	1,360	2,490	5,100	7,650	14,700	23,500	41,500	84,700
250	558	1,170	2,200	4,510	6,760	13,000	20,800	36,700	74,900
300	506	1,060	1,990	4,090	6,130	11,800	18,800	33,300	67,800
350	465	973	1,830	3,760	5,640	10,900	17,300	30,600	62,400
400	433	905	1,710	3,500	5,250	10,100	16,100	28,500	58,100
450	406	849	1,600	3,290	4,920	9,480	15,100	26,700	54,500
500	384	802	1,510	3,100	4,650	8,950	14,300	25,200	51,500
550	364	762	1,440	2,950	4,420	8,500	13,600	24,000	48,900
600	348	727	1,370	2,810	4,210	8,110	12,900	22,900	46,600
650	333	696	1,310	2,690	4,030	7,770	12,400	21,900	44,600
700	320	669	1,260	2,590	3,880	7,460	11,900	21,000	42,900
750	308	644	1,210	2,490	3,730	7,190	11,500	20,300	41,300
800	298	622	1,170	2,410	3,610	6,940	11,100	19,600	39,900
850	288	602	1,130	2,330	3,490	6,720	10,700	18,900	38,600
900	279	584	1,100	2,260	3,380	6,520	10,400	18,400	37,400
950	271	567	1,070	2,190	3,290	6,330	10,100	17,800	36,400
1,000	264	551	1,040	2,130	3,200	6,150	9,810	17,300	35,400
1,100	250	524	987	2,030	3,030	5,840	9,320	16,500	33,600
1,200	239	500	941	1,930	2,900	5,580	8,890	15,700	32,000
1,300	229	478	901	1,850	2,770	5,340	8,510	15,000	30,700
1,400	220	460	866	1,780	2,660	5,130	8,180	14,500	29,500
1,500	212	443	834	1,710	2,570	4,940	7,880	13,900	28,400
1,600	205	428	806	1,650	2,480	4,770	7,610	13,400	27,400
1,700	198	414	780	1,600	2,400	4,620	7,360	13,000	26,500
1,800	192	401	756	1,550	2,330	4,480	7,140	12,600	25,700
1,900	186	390	734	1,510	2,260	4,350	6,930	12,300	25,000
2,000	181	379	714	1,470	2,200	4,230	6,740	11,900	24,300

For SI: 1 inch = 25.4 mm, 1 foot = 304.8 mm, 1 pound per square inch = 6.895 kPa, 1-inch water column = 0.2488 kPa,
        1 British thermal unit per hour = 0.2931 W, 1 cubic foot per hour = 0.0283 m³/h, 1 degree = 0.01745 rad.
Note: All table entries have been rounded to three significant digits.

TABLE 402.4(6)
SEMIRIGID COPPER TUBING

Gas	Natural
Inlet Pressure	Less than 2 psi
Pressure Drop	0.3 in. w.c.
Specific Gravity	0.60

Nominal	K & L	$^1/_4$	$^3/_8$	$^1/_2$	$^5/_8$	$^3/_4$	1	$1^1/_4$	$1^1/_2$	2
	ACR	$^3/_8$	$^1/_2$	$^5/_8$	$^3/_4$	$^7/_8$	$1^1/_8$	$1^3/_8$	—	—
Outside		0.375	0.500	0.625	0.750	0.875	1.125	1.375	1.625	2.125
Inside		0.305	0.402	0.527	0.652	0.745	0.995	1.245	1.481	1.959
Length (ft)		\multicolumn{9}{c}{Capacity in Cubic Feet of Gas Per Hour}								
10		20	42	85	148	210	448	806	1,270	2,650
20		14	29	58	102	144	308	554	873	1,820
30		11	23	47	82	116	247	445	701	1,460
40		10	20	40	70	99	211	381	600	1,250
50		NA	17	35	62	88	187	337	532	1,110
60		NA	16	32	56	79	170	306	482	1,000
70		NA	14	29	52	73	156	281	443	924
80		NA	13	27	48	68	145	262	413	859
90		NA	13	26	45	64	136	245	387	806
100		NA	12	24	43	60	129	232	366	761
125		NA	11	22	38	53	114	206	324	675
150		NA	10	20	34	48	103	186	294	612
175		NA	NA	18	31	45	95	171	270	563
200		NA	NA	17	29	41	89	159	251	523
250		NA	NA	15	26	37	78	141	223	464
300		NA	NA	13	23	33	71	128	202	420
350		NA	NA	12	22	31	65	118	186	387
400		NA	NA	11	20	28	61	110	173	360
450		NA	NA	11	19	27	57	103	162	338
500		NA	NA	10	18	25	54	97	153	319
550		NA	NA	NA	17	24	51	92	145	303
600		NA	NA	NA	16	23	49	88	139	289
650		NA	NA	NA	15	22	47	84	133	277
700		NA	NA	NA	15	21	45	81	128	266
750		NA	NA	NA	14	20	43	78	123	256
800		NA	NA	NA	14	20	42	75	119	247
850		NA	NA	NA	13	19	40	73	115	239
900		NA	NA	NA	13	18	39	71	111	232
950		NA	NA	NA	13	18	38	69	108	225
1,000		NA	NA	NA	12	17	37	67	105	219
1,100		NA	NA	NA	12	16	35	63	100	208
1,200		NA	NA	NA	11	16	34	60	95	199
1,300		NA	NA	NA	11	15	32	58	91	190
1,400		NA	NA	NA	10	14	31	56	88	183
1,500		NA	NA	NA	NA	14	30	54	84	176
1,600		NA	NA	NA	NA	13	29	52	82	170
1,700		NA	NA	NA	NA	13	28	50	79	164
1,800		NA	NA	NA	NA	13	27	49	77	159
1,900		NA	NA	NA	NA	12	26	47	74	155
2,000		NA	NA	NA	NA	12	25	46	72	151

For SI: 1 inch = 25.4 mm, 1 foot = 304.8 mm, 1 pound per square inch = 6.895 kPa, 1-inch water column = 0.2488 kPa,
1 British thermal unit per hour = 0.2931 W, 1 cubic foot per hour = 0.0283 m³/h, 1 degree = 0.01745 rad.

Notes:

1. Table capacities are based on Type K copper tubing inside diameter (shown), which has the smallest inside diameter of the copper tubing products.

2. NA means a flow of less than 10 cfh.

3. All table entries have been rounded to three significant digits.

TABLE 402.4(7)
SEMIRIGID COPPER TUBING

Gas	Natural
Inlet Pressure	Less than 2 psi
Pressure Drop	0.5 in. w.c.
Specific Gravity	0.60

Nominal		TUBE SIZE (inch)								
Nominal	K & L	$^1/_4$	$^3/_8$	$^1/_2$	$^5/_8$	$^3/_4$	1	$1^1/_4$	$1^1/_2$	2
	ACR	$^3/_8$	$^1/_2$	$^5/_8$	$^3/_4$	$^7/_8$	$1^1/_8$	$1^3/_8$	—	—
Outside		0.375	0.500	0.625	0.750	0.875	1.125	1.375	1.625	2.125
Inside		0.305	0.402	0.527	0.652	0.745	0.995	1.245	1.481	1.959
Length (ft)		Capacity in Cubic Feet of Gas Per Hour								
10		27	55	111	195	276	590	1,060	1,680	3,490
20		18	38	77	134	190	406	730	1,150	2,400
30		15	30	61	107	152	326	586	925	1,930
40		13	26	53	92	131	279	502	791	1,650
50		11	23	47	82	116	247	445	701	1,460
60		10	21	42	74	105	224	403	635	1,320
70		NA	19	39	68	96	206	371	585	1,220
80		NA	18	36	63	90	192	345	544	1,130
90		NA	17	34	59	84	180	324	510	1,060
100		NA	16	32	56	79	170	306	482	1,000
125		NA	14	28	50	70	151	271	427	890
150		NA	13	26	45	64	136	245	387	806
175		NA	12	24	41	59	125	226	356	742
200		NA	11	22	39	55	117	210	331	690
250		NA	NA	20	34	48	103	186	294	612
300		NA	NA	18	31	44	94	169	266	554
350		NA	NA	16	28	40	86	155	245	510
400		NA	NA	15	26	38	80	144	228	474
450		NA	NA	14	25	35	75	135	214	445
500		NA	NA	13	23	33	71	128	202	420
550		NA	NA	13	22	32	68	122	192	399
600		NA	NA	12	21	30	64	116	183	381
650		NA	NA	12	20	29	62	111	175	365
700		NA	NA	11	20	28	59	107	168	350
750		NA	NA	11	19	27	57	103	162	338
800		NA	NA	10	18	26	55	99	156	326
850		NA	NA	10	18	25	53	96	151	315
900		NA	NA	NA	17	24	52	93	147	306
950		NA	NA	NA	17	24	50	90	143	297
1,000		NA	NA	NA	16	23	49	88	139	289
1,100		NA	NA	NA	15	22	46	84	132	274
1,200		NA	NA	NA	15	21	44	80	126	262
1,300		NA	NA	NA	14	20	42	76	120	251
1,400		NA	NA	NA	13	19	41	73	116	241
1,500		NA	NA	NA	13	18	39	71	111	232
1,600		NA	NA	NA	13	18	38	68	108	224
1,700		NA	NA	NA	12	17	37	66	104	217
1,800		NA	NA	NA	12	17	36	64	101	210
1,900		NA	NA	NA	11	16	35	62	98	204
2,000		NA	NA	NA	11	16	34	60	95	199

For SI: 1 inch = 25.4 mm, 1 foot = 304.8 mm, 1 pound per square inch = 6.895 kPa, 1-inch water column = 0.2488 kPa,
1 British thermal unit per hour = 0.2931 W, 1 cubic foot per hour = 0.0283 m³/h, 1 degree = 0.01745 rad.

Notes:
1. Table capacities are based on Type K copper tubing inside diameter (shown), which has the smallest inside diameter of the copper tubing products.
2. NA means a flow of less than 10 cfh.
3. All table entries have been rounded to three significant digits.

**TABLE 402.4(8)
SEMIRIGID COPPER TUBING**

Gas	Natural
Inlet Pressure	Less than 2 psi
Pressure Drop	1.0 in. w.c.
Specific Gravity	0.60

						TUBE SIZE (inch)				
Nominal	**K & L**	$^1/_4$	$^3/_8$	$^1/_2$	$^5/_8$	$^3/_4$	1	$1^1/_4$	$1^1/_2$	2
	ACR	$^3/_8$	$^1/_2$	$^5/_8$	$^3/_4$	$^7/_8$	$1^1/_8$	$1^3/_8$	—	—
Outside		0.375	0.500	0.625	0.750	0.875	1.125	1.375	1.625	2.125
Inside		0.305	0.402	0.527	0.652	0.745	0.995	1.245	1.481	1.959
Length (ft)		Capacity in Cubic Feet of Gas Per Hour								
10		39	80	162	283	402	859	1,550	2,440	5,080
20		27	55	111	195	276	590	1,060	1,680	3,490
30		21	44	89	156	222	474	853	1,350	2,800
40		18	38	77	134	190	406	730	1,150	2,400
50		16	33	68	119	168	359	647	1,020	2,130
60		15	30	61	107	152	326	586	925	1,930
70		13	28	57	99	140	300	539	851	1,770
80		13	26	53	92	131	279	502	791	1,650
90		12	24	49	86	122	262	471	742	1,550
100		11	23	47	82	116	247	445	701	1,460
125		NA	20	41	72	103	219	394	622	1,290
150		NA	18	37	65	93	198	357	563	1,170
175		NA	17	34	60	85	183	329	518	1,080
200		NA	16	32	56	79	170	306	482	1,000
250		NA	14	28	50	70	151	271	427	890
300		NA	13	26	45	64	136	245	387	806
350		NA	12	24	41	59	125	226	356	742
400		NA	11	22	39	55	117	210	331	690
450		NA	10	21	36	51	110	197	311	647
500		NA	NA	20	34	48	103	186	294	612
550		NA	NA	19	32	46	98	177	279	581
600		NA	NA	18	31	44	94	169	266	554
650		NA	NA	17	30	42	90	162	255	531
700		NA	NA	16	28	40	86	155	245	510
750		NA	NA	16	27	39	83	150	236	491
800		NA	NA	15	26	38	80	144	228	474
850		NA	NA	15	26	36	78	140	220	459
900		NA	NA	14	25	35	75	135	214	445
950		NA	NA	14	24	34	73	132	207	432
1,000		NA	NA	13	23	33	71	128	202	420
1,100		NA	NA	13	22	32	68	122	192	399
1,200		NA	NA	12	21	30	64	116	183	381
1,300		NA	NA	12	20	29	62	111	175	365
1,400		NA	NA	11	20	28	59	107	168	350
1,500		NA	NA	11	19	27	57	103	162	338
1,600		NA	NA	10	18	26	55	99	156	326
1,700		NA	NA	10	18	25	53	96	151	315
1,800		NA	NA	NA	17	24	52	93	147	306
1,900		NA	NA	NA	17	24	50	90	143	297
2,000		NA	NA	NA	16	23	49	88	139	289

For SI: 1 inch = 25.4 mm, 1 foot = 304.8 mm, 1 pound per square inch = 6.895 kPa, 1-inch water column = 0.2488 kPa,
1 British thermal unit per hour = 0.2931 W, 1 cubic foot per hour = 0.0283 m³/h, 1 degree = 0.01745 rad.

Notes:

1. Table capacities are based on Type K copper tubing inside diameter (shown), which has the smallest inside diameter of the copper tubing products.
2. NA means a flow of less than 10 cfh.
3. All table entries have been rounded to three significant digits.

TABLE 402.4(9)
SEMIRIGID COPPER TUBING

Gas	Natural
Inlet Pressure	Less than 2.0 psi
Pressure Drop	17.0 in w.c.
Specific Gravity	0.60

					TUBE SIZE (inch)						
Nominal	**K & L**	$^1/_4$	$^3/_8$	$^1/_2$	$^5/_8$	$^3/_4$	1	$1^1/_4$	$1^1/_2$	2
	ACR	$^3/_8$	$^1/_2$	$^5/_8$	$^3/_4$	$^7/_8$	$1^1/_8$	$1^3/_8$	—	—
Outside		0.375	0.500	0.625	0.750	0.875	1.125	1.375	1.625	2.125
Inside		0.305	0.402	0.527	0.652	0.745	0.995	1.245	1.481	1.959
Length (ft)		Capacity in Cubic Feet of Gas Per Hour								
10		190	391	796	1,390	1,970	4,220	7,590	12,000	24,900
20		130	269	547	956	1,360	2,900	5,220	8,230	17,100
30		105	216	439	768	1,090	2,330	4,190	6,610	13,800
40		90	185	376	657	932	1,990	3,590	5,650	11,800
50		79	164	333	582	826	1,770	3,180	5,010	10,400
60		72	148	302	528	749	1,600	2,880	4,540	9,460
70		66	137	278	486	689	1,470	2,650	4,180	8,700
80		62	127	258	452	641	1,370	2,460	3,890	8,090
90		58	119	243	424	601	1,280	2,310	3,650	7,590
100		55	113	229	400	568	1,210	2,180	3,440	7,170
125		48	100	203	355	503	1,080	1,940	3,050	6,360
150		44	90	184	321	456	974	1,750	2,770	5,760
175		40	83	169	296	420	896	1,610	2,540	5,300
200		38	77	157	275	390	834	1,500	2,370	4,930
250		33	69	140	244	346	739	1,330	2,100	4,370
300		30	62	126	221	313	670	1,210	1,900	3,960
350		28	57	116	203	288	616	1,110	1,750	3,640
400		26	53	108	189	268	573	1,030	1,630	3,390
450		24	50	102	177	252	538	968	1,530	3,180
500		23	47	96	168	238	508	914	1,440	3,000
550		22	45	91	159	226	482	868	1,370	2,850
600		21	43	87	152	215	460	829	1,310	2,720
650		20	41	83	145	206	441	793	1,250	2,610
700		19	39	80	140	198	423	762	1,200	2,500
750		18	38	77	135	191	408	734	1,160	2,410
800		18	37	74	130	184	394	709	1,120	2,330
850		17	35	72	126	178	381	686	1,080	2,250
900		17	34	70	122	173	370	665	1,050	2,180
950		16	33	68	118	168	359	646	1,020	2,120
1,000		16	32	66	115	163	349	628	991	2,060
1,100		15	31	63	109	155	332	597	941	1,960
1,200		14	29	60	104	148	316	569	898	1,870
1,300		14	28	57	100	142	303	545	860	1,790
1,400		13	27	55	96	136	291	524	826	1,720
1,500		13	26	53	93	131	280	505	796	1,660
1,600		12	25	51	89	127	271	487	768	1,600
1,700		12	24	49	86	123	262	472	744	1,550
1,800		11	24	48	84	119	254	457	721	1,500
1,900		11	23	47	81	115	247	444	700	1,460
2,000		11	22	45	79	112	240	432	681	1,420

For SI: 1 inch = 25.4 mm, 1 foot = 304.8 mm, 1 pound per square inch = 6.895 kPa, 1-inch water column = 0.2488 kPa, 1 British thermal unit per hour = 0.2931 W, 1 cubic foot per hour = 0.0283 m³/h, 1 degree = 0.01745 rad.

Notes:
1. Table capacities are based on Type K copper tubing inside diameter (shown), which has the smallest inside diameter of the copper tubing products.
2. All table entries have been rounded to three significant digits.

		Gas	Natural
TABLE 402.4(10)		**Inlet Pressure**	2.0 psi
SEMIRIGID COPPER TUBING		**Pressure Drop**	1.0 psi
		Specific Gravity	0.60

		TUBE SIZE (inch)								
Nominal	**K & L**	$1/4$	$3/8$	$1/2$	$5/8$	$3/4$	1	$1^1/4$	$1^1/2$	2
	ACR	$3/8$	$1/2$	$5/8$	$3/4$	$7/8$	$1^1/8$	$1^3/8$	—	—
Outside		0.375	0.500	0.625	0.750	0.875	1.125	1.375	1.625	2.125
Inside		0.305	0.402	0.527	0.652	0.745	0.995	1.245	1.481	1.959
Length (ft)		Capacity in Cubic Feet of Gas Per Hour								
10		245	506	1,030	1,800	2,550	5,450	9,820	15,500	32,200
20		169	348	708	1,240	1,760	3,750	6,750	10,600	22,200
30		135	279	568	993	1,410	3,010	5,420	8,550	17,800
40		116	239	486	850	1,210	2,580	4,640	7,310	15,200
50		103	212	431	754	1,070	2,280	4,110	6,480	13,500
60		93	192	391	683	969	2,070	3,730	5,870	12,200
70		86	177	359	628	891	1,900	3,430	5,400	11,300
80		80	164	334	584	829	1,770	3,190	5,030	10,500
90		75	154	314	548	778	1,660	2,990	4,720	9,820
100		71	146	296	518	735	1,570	2,830	4,450	9,280
125		63	129	263	459	651	1,390	2,500	3,950	8,220
150		57	117	238	416	590	1,260	2,270	3,580	7,450
175		52	108	219	383	543	1,160	2,090	3,290	6,850
200		49	100	204	356	505	1,080	1,940	3,060	6,380
250		43	89	181	315	448	956	1,720	2,710	5,650
300		39	80	164	286	406	866	1,560	2,460	5,120
350		36	74	150	263	373	797	1,430	2,260	4,710
400		33	69	140	245	347	741	1,330	2,100	4,380
450		31	65	131	230	326	696	1,250	1,970	4,110
500		30	61	124	217	308	657	1,180	1,870	3,880
550		28	58	118	206	292	624	1,120	1,770	3,690
600		27	55	112	196	279	595	1,070	1,690	3,520
650		26	53	108	188	267	570	1,030	1,620	3,370
700		25	51	103	181	256	548	986	1,550	3,240
750		24	49	100	174	247	528	950	1,500	3,120
800		23	47	96	168	239	510	917	1,450	3,010
850		22	46	93	163	231	493	888	1,400	2,920
900		22	44	90	158	224	478	861	1,360	2,830
950		21	43	88	153	217	464	836	1,320	2,740
1,000		20	42	85	149	211	452	813	1,280	2,670
1,100		19	40	81	142	201	429	772	1,220	2,540
1,200		18	38	77	135	192	409	737	1,160	2,420
1,300		18	36	74	129	183	392	705	1,110	2,320
1,400		17	35	71	124	176	376	678	1,070	2,230
1,500		16	34	68	120	170	363	653	1,030	2,140
1,600		16	33	66	116	164	350	630	994	2,070
1,700		15	31	64	112	159	339	610	962	2,000
1,800		15	30	62	108	154	329	592	933	1,940
1,900		14	30	60	105	149	319	575	906	1,890
2,000		14	29	59	102	145	310	559	881	1,830

For SI: 1 inch = 25.4 mm, 1 foot = 304.8 mm, 1 pound per square inch = 6.895 kPa, 1-inch water column = 0.2488 kPa,
1 British thermal unit per hour = 0.2931 W, 1 cubic foot per hour = 0.0283 m³/h, 1 degree = 0.01745 rad.

Notes:

1. Table capacities are based on Type K copper tubing inside diameter (shown), which has the smallest inside diameter of the copper tubing products.
2. All table entries have been rounded to three significant digits.

	Gas	Natural
	Inlet Pressure	2.0 psi
	Pressure Drop	1.5 psi
	Specific Gravity	0.60

TABLE 402.4(11)
SEMIRIGID COPPER TUBING

SPECIAL USE	Pipe sizing between point of delivery and the house line regulator. Total load supplied by a single house line regulator not exceeding 150 cubic feet per hour.								
TUBE SIZE (inch)									
Nominal	K & L	$^1/_4$	$^3/_8$	$^1/_2$	$^5/_8$	$^3/_4$	1	$1^1/_4$	$1^1/_2$	2
	ACR	$^3/_8$	$^1/_2$	$^5/_8$	$^3/_4$	$^7/_8$	$1^1/_8$	$1^3/_8$	—	—
Outside		0.375	0.500	0.625	0.750	0.875	1.125	1.375	1.625	2.125
Inside		0.305	0.402	0.527	0.652	0.745	0.995	1.245	1.481	1.959
Length (ft)		**Capacity in Cubic Feet of Gas Per Hour**								
10		303	625	1,270	2,220	3,150	6,740	12,100	19,100	39,800
20		208	430	874	1,530	2,170	4,630	8,330	13,100	27,400
30		167	345	702	1,230	1,740	3,720	6,690	10,600	22,000
40		143	295	601	1,050	1,490	3,180	5,730	9,030	18,800
50		127	262	532	931	1,320	2,820	5,080	8,000	16,700
60		115	237	482	843	1,200	2,560	4,600	7,250	15,100
70		106	218	444	776	1,100	2,350	4,230	6,670	13,900
80		98	203	413	722	1,020	2,190	3,940	6,210	12,900
90		92	190	387	677	961	2,050	3,690	5,820	12,100
100		87	180	366	640	907	1,940	3,490	5,500	11,500
125		77	159	324	567	804	1,720	3,090	4,880	10,200
150		70	144	294	514	729	1,560	2,800	4,420	9,200
175		64	133	270	472	670	1,430	2,580	4,060	8,460
200		60	124	252	440	624	1,330	2,400	3,780	7,870
250		53	110	223	390	553	1,180	2,130	3,350	6,980
300		48	99	202	353	501	1,070	1,930	3,040	6,320
350		44	91	186	325	461	984	1,770	2,790	5,820
400		41	85	173	302	429	916	1,650	2,600	5,410
450		39	80	162	283	402	859	1,550	2,440	5,080
500		36	75	153	268	380	811	1,460	2,300	4,800
550		35	72	146	254	361	771	1,390	2,190	4,560
600		33	68	139	243	344	735	1,320	2,090	4,350
650		32	65	133	232	330	704	1,270	2,000	4,160
700		30	63	128	223	317	676	1,220	1,920	4,000
750		29	60	123	215	305	652	1,170	1,850	3,850
800		28	58	119	208	295	629	1,130	1,790	3,720
850		27	57	115	201	285	609	1,100	1,730	3,600
900		27	55	111	195	276	590	1,060	1,680	3,490
950		26	53	108	189	268	573	1,030	1,630	3,390
1,000		25	52	105	184	261	558	1,000	1,580	3,300
1,100		24	49	100	175	248	530	954	1,500	3,130
1,200		23	47	95	167	237	505	910	1,430	2,990
1,300		22	45	91	160	227	484	871	1,370	2,860
1,400		21	43	88	153	218	465	837	1,320	2,750
1,500		20	42	85	148	210	448	806	1,270	2,650
1,600		19	40	82	143	202	432	779	1,230	2,560
1,700		19	39	79	138	196	419	753	1,190	2,470
1,800		18	38	77	134	190	406	731	1,150	2,400
1,900		18	37	74	130	184	394	709	1,120	2,330
2,000		17	36	72	126	179	383	690	1,090	2,270

For SI: 1 inch = 25.4 mm, 1 foot = 304.8 mm, 1 pound per square inch = 6.895 kPa, 1-inch water column = 0.2488 kPa, 1 British thermal unit per hour = 0.2931 W, 1 cubic foot per hour = 0.0283 m³/h, 1 degree = 0.01745 rad.

Notes:
1. Table capacities are based on Type K copper tubing inside diameter (shown), which has the smallest inside diameter of the copper tubing products.
2. Where this table is used to size the tubing upstream of a line pressure regulator, the pipe or tubing downstream of the line pressure regulator shall be sized using a pressure drop not greater than 1 inch w.c.
3. All table entries have been rounded to three significant digits.

TABLE 402.4(12)
SEMIRIGID COPPER TUBING

Gas	Natural
Inlet Pressure	5.0 psi
Pressure Drop	3.5 psi
Specific Gravity	0.60

Nominal	K & L	\(^1/_4\)	\(^3/_8\)	\(^1/_2\)	\(^5/_8\)	\(^3/_4\)	1	\(1^1/_4\)	\(1^1/_2\)	2
	ACR	\(^3/_8\)	\(^1/_2\)	\(^5/_8\)	\(^3/_4\)	\(^7/_8\)	\(1^1/_8\)	\(1^3/_8\)	—	—
Outside		0.375	0.500	0.625	0.750	0.875	1.125	1.375	1.625	2.125
Inside		0.305	0.402	0.527	0.652	0.745	0.995	1.245	1.481	1.959
Length (ft)		Capacity in Cubic Feet of Gas Per Hour								
10		511	1,050	2,140	3,750	5,320	11,400	20,400	32,200	67,100
20		351	724	1,470	2,580	3,650	7,800	14,000	22,200	46,100
30		282	582	1,180	2,070	2,930	6,270	11,300	17,800	37,000
40		241	498	1,010	1,770	2,510	5,360	9,660	15,200	31,700
50		214	441	898	1,570	2,230	4,750	8,560	13,500	28,100
60		194	400	813	1,420	2,020	4,310	7,750	12,200	25,500
70		178	368	748	1,310	1,860	3,960	7,130	11,200	23,400
80		166	342	696	1,220	1,730	3,690	6,640	10,500	21,800
90		156	321	653	1,140	1,620	3,460	6,230	9,820	20,400
100		147	303	617	1,080	1,530	3,270	5,880	9,270	19,300
125		130	269	547	955	1,360	2,900	5,210	8,220	17,100
150		118	243	495	866	1,230	2,620	4,720	7,450	15,500
175		109	224	456	796	1,130	2,410	4,350	6,850	14,300
200		101	208	424	741	1,050	2,250	4,040	6,370	13,300
250		90	185	376	657	932	1,990	3,580	5,650	11,800
300		81	167	340	595	844	1,800	3,250	5,120	10,700
350		75	154	313	547	777	1,660	2,990	4,710	9,810
400		69	143	291	509	722	1,540	2,780	4,380	9,120
450		65	134	273	478	678	1,450	2,610	4,110	8,560
500		62	127	258	451	640	1,370	2,460	3,880	8,090
550		58	121	245	429	608	1,300	2,340	3,690	7,680
600		56	115	234	409	580	1,240	2,230	3,520	7,330
650		53	110	224	392	556	1,190	2,140	3,370	7,020
700		51	106	215	376	534	1,140	2,050	3,240	6,740
750		49	102	207	362	514	1,100	1,980	3,120	6,490
800		48	98	200	350	497	1,060	1,910	3,010	6,270
850		46	95	194	339	481	1,030	1,850	2,910	6,070
900		45	92	188	328	466	1,000	1,790	2,820	5,880
950		43	90	182	319	452	967	1,740	2,740	5,710
1,000		42	87	177	310	440	940	1,690	2,670	5,560
1,100		40	83	169	295	418	893	1,610	2,530	5,280
1,200		38	79	161	281	399	852	1,530	2,420	5,040
1,300		37	76	154	269	382	816	1,470	2,320	4,820
1,400		35	73	148	259	367	784	1,410	2,220	4,630
1,500		34	70	143	249	353	755	1,360	2,140	4,460
1,600		33	68	138	241	341	729	1,310	2,070	4,310
1,700		32	65	133	233	330	705	1,270	2,000	4,170
1,800		31	63	129	226	320	684	1,230	1,940	4,040
1,900		30	62	125	219	311	664	1,200	1,890	3,930
2,000		29	60	122	213	302	646	1,160	1,830	3,820

For SI: 1 inch = 25.4 mm, 1 foot = 304.8 mm, 1 pound per square inch = 6.895 kPa, 1-inch water column = 0.2488 kPa,
1 British thermal unit per hour = 0.2931 W, 1 cubic foot per hour = 0.0283 m³/h, 1 degree = 0.01745 rad.

Notes:
1. Table capacities are based on Type K copper tubing inside diameter (shown), which has the smallest inside diameter of the copper tubing products.
2. All table entries have been rounded to three significant digits.

TABLE 402.4(13)
CORRUGATED STAINLESS STEEL TUBING (CSST)

Gas	Natural
Inlet Pressure	Less than 2 psi
Pressure Drop	0.5 in. w.c.
Specific Gravity	0.60

	TUBE SIZE (EHD)												
Flow Designation	13	15	18	19	23	25	30	31	37	46	48	60	62
Length (ft)	Capacity in Cubic Feet of Gas Per Hour												
5	46	63	115	134	225	270	471	546	895	1,790	2,070	3,660	4,140
10	32	44	82	95	161	192	330	383	639	1,260	1,470	2,600	2,930
15	25	35	66	77	132	157	267	310	524	1,030	1,200	2,140	2,400
20	22	31	58	67	116	137	231	269	456	888	1,050	1,850	2,080
25	19	27	52	60	104	122	206	240	409	793	936	1,660	1,860
30	18	25	47	55	96	112	188	218	374	723	856	1,520	1,700
40	15	21	41	47	83	97	162	188	325	625	742	1,320	1,470
50	13	19	37	42	75	87	144	168	292	559	665	1,180	1,320
60	12	17	34	38	68	80	131	153	267	509	608	1,080	1,200
70	11	16	31	36	63	74	121	141	248	471	563	1,000	1,110
80	10	15	29	33	60	69	113	132	232	440	527	940	1,040
90	10	14	28	32	57	65	107	125	219	415	498	887	983
100	9	13	26	30	54	62	101	118	208	393	472	843	933
150	7	10	20	23	42	48	78	91	171	320	387	691	762
200	6	9	18	21	38	44	71	82	148	277	336	600	661
250	5	8	16	19	34	39	63	74	133	247	301	538	591
300	5	7	15	17	32	36	57	67	95	226	275	492	540

For SI: 1 inch = 25.4 mm, 1 foot = 304.8 mm, 1 pound per square inch = 6.895 kPa, 1-inch water column = 0.2488 kPa,
1 British thermal unit per hour = 0.2931 W, 1 cubic foot per hour = 0.0283 m³/h, 1 degree = 0.01745 rad.

Notes:

1. Table includes losses for four 90-degree bends and two end fittings. Tubing runs with larger numbers of bends and/or fittings shall be increased by an equivalent length of tubing to the following equation: $L = 1.3n$, where L is additional length (feet) of tubing and n is the number of additional fittings and/or bends.

2. EHD—Equivalent Hydraulic Diameter, which is a measure of the relative hydraulic efficiency between different tubing sizes. The greater the value of EHD, the greater the gas capacity of the tubing.

3. All table entries have been rounded to three significant digits.

TABLE 402.4(14)
CORRUGATED STAINLESS STEEL TUBING (CSST)

Gas	Natural
Inlet Pressure	Less than 2 psi
Pressure Drop	3.0 in. w.c.
Specific Gravity	0.60

TUBE SIZE (EHD)													
Flow Designation	13	15	18	19	23	25	30	31	37	46	48	60	62
Length (ft)	Capacity in Cubic Feet of Gas Per Hour												
5	120	160	277	327	529	649	1,180	1,370	2,140	4,430	5,010	8,800	10,100
10	83	112	197	231	380	462	828	958	1,530	3,200	3,560	6,270	7,160
15	67	90	161	189	313	379	673	778	1,250	2,540	2,910	5,140	5,850
20	57	78	140	164	273	329	580	672	1,090	2,200	2,530	4,460	5,070
25	51	69	125	147	245	295	518	599	978	1,960	2,270	4,000	4,540
30	46	63	115	134	225	270	471	546	895	1,790	2,070	3,660	4,140
40	39	54	100	116	196	234	407	471	778	1,550	1,800	3,180	3,590
50	35	48	89	104	176	210	363	421	698	1,380	1,610	2,850	3,210
60	32	44	82	95	161	192	330	383	639	1,260	1,470	2,600	2,930
70	29	41	76	88	150	178	306	355	593	1,170	1,360	2,420	2,720
80	27	38	71	82	141	167	285	331	555	1,090	1,280	2,260	2,540
90	26	36	67	77	133	157	268	311	524	1,030	1,200	2,140	2,400
100	24	34	63	73	126	149	254	295	498	974	1,140	2,030	2,280
150	19	27	52	60	104	122	206	240	409	793	936	1,660	1,860
200	17	23	45	52	91	106	178	207	355	686	812	1,440	1,610
250	15	21	40	46	82	95	159	184	319	613	728	1,290	1,440
300	13	19	37	42	75	87	144	168	234	559	665	1,180	1,320

For SI: 1 inch = 25.4 mm, 1 foot = 304.8 mm, 1 pound per square inch = 6.895 kPa, 1-inch water column = 0.2488 kPa,
1 British thermal unit per hour = 0.2931 W, 1 cubic foot per hour = 0.0283 m³/h, 1 degree = 0.01745 rad.

Notes:

1. Table includes losses for four 90-degree bends and two end fittings. Tubing runs with larger numbers of bends and/or fittings shall be increased by an equivalent length of tubing to the following equation: $L = 1.3n$ where L is additional length (feet) of tubing and n is the number of additional fittings and/or bends.

2. EHD—Equivalent Hydraulic Diameter, which is a measure of the relative hydraulic efficiency between different tubing sizes. The greater the value of EHD, the greater the gas capacity of the tubing.

3. All table entries have been rounded to three significant digits.

TABLE 402.4(15)
CORRUGATED STAINLESS STEEL TUBING (CSST)

Gas	Natural
Inlet Pressure	Less than 2 psi
Pressure Drop	6.0 in. w.c.
Specific Gravity	0.60

	TUBE SIZE (EHD)												
Flow Designation	13	15	18	19	23	25	30	31	37	46	48	60	62
Length (ft)	Capacity in Cubic Feet of Gas Per Hour												
5	173	229	389	461	737	911	1,690	1,950	3,000	6,280	7,050	12,400	14,260
10	120	160	277	327	529	649	1,180	1,370	2,140	4,430	5,010	8,800	10,100
15	96	130	227	267	436	532	960	1,110	1,760	3,610	4,100	7,210	8,260
20	83	112	197	231	380	462	828	958	1,530	3,120	3,560	6,270	7,160
25	74	99	176	207	342	414	739	855	1,370	2,790	3,190	5,620	6,400
30	67	90	161	189	313	379	673	778	1,250	2,540	2,910	5,140	5,850
40	57	78	140	164	273	329	580	672	1,090	2,200	2,530	4,460	5,070
50	51	69	125	147	245	295	518	599	978	1,960	2,270	4,000	4,540
60	46	63	115	134	225	270	471	546	895	1,790	2,070	3,660	4,140
70	42	58	106	124	209	250	435	505	830	1,660	1,920	3,390	3,840
80	39	54	100	116	196	234	407	471	778	1,550	1,800	3,180	3,590
90	37	51	94	109	185	221	383	444	735	1,460	1,700	3,000	3,390
100	35	48	89	104	176	210	363	421	698	1,380	1,610	2,850	3,210
150	28	39	73	85	145	172	294	342	573	1,130	1,320	2,340	2,630
200	24	34	63	73	126	149	254	295	498	974	1,140	2,030	2,280
250	21	30	57	66	114	134	226	263	447	870	1,020	1,820	2,040
300	19	27	52	60	104	122	206	240	409	793	936	1,660	1,860

For SI: 1 inch = 25.4 mm, 1 foot = 304.8 mm, 1 pound per square inch = 6.895 kPa, 1-inch water column = 0.2488 kPa,
1 British thermal unit per hour = 0.2931 W, 1 cubic foot per hour = 0.0283 m³/h, 1 degree = 0.01745 rad.

Notes:

1. Table includes losses for four 90-degree bends and two end fittings. Tubing runs with larger numbers of bends and/or fittings shall be increased by an equivalent length of tubing to the following equation: $L = 1.3n$ where L is additional length (feet) of tubing and n is the number of additional fittings and/or bends.
2. EHD—Equivalent Hydraulic Diameter, which is a measure of the relative hydraulic efficiency between different tubing sizes. The greater the value of EHD, the greater the gas capacity of the tubing.
3. All table entries have been rounded to three significant digits.

TABLE 402.4(16)
CORRUGATED STAINLESS STEEL TUBING (CSST)

Gas	Natural
Inlet Pressure	2.0 psi
Pressure Drop	1.0 psi
Specific Gravity	0.60

	TUBE SIZE (EHD)												
Flow Designation	13	15	18	19	23	25	30	31	37	46	48	60	62
Length (ft)	Capacity in Cubic Feet of Gas Per Hour												
10	270	353	587	700	1,100	1,370	2,590	2,990	4,510	9,600	10,700	18,600	21,600
25	166	220	374	444	709	876	1,620	1,870	2,890	6,040	6,780	11,900	13,700
30	151	200	342	405	650	801	1,480	1,700	2,640	5,510	6,200	10,900	12,500
40	129	172	297	351	567	696	1,270	1,470	2,300	4,760	5,380	9,440	10,900
50	115	154	266	314	510	624	1,140	1,310	2,060	4,260	4,820	8,470	9,720
75	93	124	218	257	420	512	922	1,070	1,690	3,470	3,950	6,940	7,940
80	89	120	211	249	407	496	892	1,030	1,640	3,360	3,820	6,730	7,690
100	79	107	189	222	366	445	795	920	1,470	3,000	3,420	6,030	6,880
150	64	87	155	182	302	364	646	748	1,210	2,440	2,800	4,940	5,620
200	55	75	135	157	263	317	557	645	1,050	2,110	2,430	4,290	4,870
250	49	67	121	141	236	284	497	576	941	1,890	2,180	3,850	4,360
300	44	61	110	129	217	260	453	525	862	1,720	1,990	3,520	3,980
400	38	52	96	111	189	225	390	453	749	1,490	1,730	3,060	3,450
500	34	46	86	100	170	202	348	404	552	1,330	1,550	2,740	3,090

For SI: 1 inch = 25.4 mm, 1 foot = 304.8 mm, 1 pound per square inch = 6.895 kPa, 1-inch water column = 0.2488 kPa,
1 British thermal unit per hour = 0.2931 W, 1 cubic foot per hour = 0.0283 m^3/h, 1 degree = 0.01745 rad.

Notes:

1. Table does not include effect of pressure drop across the line regulator. Where regulator loss exceeds $^3/_4$ psi, DO NOT USE THIS TABLE. Consult with the regulator manufacturer for pressure drops and capacity factors. Pressure drops across a regulator may vary with flow rate.

2. CAUTION: Capacities shown in the table might exceed maximum capacity for a selected regulator. Consult with the regulator or tubing manufacturer for guidance.

3. Table includes losses for four 90-degree bends and two end fittings. Tubing runs with larger numbers of bends and/or fittings shall be increased by an equivalent length of tubing to the following equation: $L = 1.3n$ where L is additional length (feet) of tubing and n is the number of additional fittings and/or bends.

4. EHD—Equivalent Hydraulic Diameter, which is a measure of the relative hydraulic efficiency between different tubing sizes. The greater the value of EHD, the greater the gas capacity of the tubing.

5. All table entries have been rounded to three significant digits.

TABLE 402.4(17)
CORRUGATED STAINLESS STEEL TUBING (CSST)

Gas	Natural
Inlet Pressure	5.0 psi
Pressure Drop	3.5 psi
Specific Gravity	0.60

	TUBE SIZE (EHD)												
Flow Designation	13	15	18	19	23	25	30	31	37	46	48	60	62
Length (ft)	Capacity in Cubic Feet of Gas Per Hour												
10	523	674	1,080	1,300	2,000	2,530	4,920	5,660	8,300	18,100	19,800	34,400	40,400
25	322	420	691	827	1,290	1,620	3,080	3,540	5,310	11,400	12,600	22,000	25,600
30	292	382	632	755	1,180	1,480	2,800	3,230	4,860	10,400	11,500	20,100	23,400
40	251	329	549	654	1,030	1,280	2,420	2,790	4,230	8,970	10,000	17,400	20,200
50	223	293	492	586	926	1,150	2,160	2,490	3,790	8,020	8,930	15,600	18,100
75	180	238	403	479	763	944	1,750	2,020	3,110	6,530	7,320	12,800	14,800
80	174	230	391	463	740	915	1,690	1,960	3,020	6,320	7,090	12,400	14,300
100	154	205	350	415	665	820	1,510	1,740	2,710	5,650	6,350	11,100	12,800
150	124	166	287	339	548	672	1,230	1,420	2,220	4,600	5,200	9,130	10,500
200	107	143	249	294	478	584	1,060	1,220	1,930	3,980	4,510	7,930	9,090
250	95	128	223	263	430	524	945	1,090	1,730	3,550	4,040	7,110	8,140
300	86	116	204	240	394	479	860	995	1,590	3,240	3,690	6,500	7,430
400	74	100	177	208	343	416	742	858	1,380	2,800	3,210	5,650	6,440
500	66	89	159	186	309	373	662	766	1,040	2,500	2,870	5,060	5,760

For SI: 1 inch = 25.4 mm, 1 foot = 304.8 mm, 1 pound per square inch = 6.895 kPa, 1-inch water column = 0.2488 kPa,
1 British thermal unit per hour = 0.2931 W, 1 cubic foot per hour = 0.0283 m³/h, 1 degree = 0.01745 rad.

Notes:

1. Table does not include effect of pressure drop across the line regulator. Where regulator loss exceeds $^3/_4$ psi, DO NOT USE THIS TABLE. Consult with the regulator manufacturer for pressure drops and capacity factors. Pressure drops across a regulator may vary with flow rate.

2. CAUTION: Capacities shown in the table might exceed maximum capacity for a selected regulator. Consult with the regulator or tubing manufacturer for guidance.

3. Table includes losses for four 90-degree bends and two end fittings. Tubing runs with larger numbers of bends and/or fittings shall be increased by an equivalent length of tubing to the following equation: $L = 1.3n$ where L is additional length (feet) of tubing and n is the number of additional fittings and/or bends.

4. EHD—Equivalent Hydraulic Diameter, which is a measure of the relative hydraulic efficiency between different tubing sizes. The greater the value of EHD, the greater the gas capacity of the tubing.

5. All table entries have been rounded to three significant digits.

TABLE 402.4(18)
POLYETHYLENE PLASTIC PIPE

Gas	Natural
Inlet Pressure	Less than 2 psi
Pressure Drop	0.3 in. w.c.
Specific Gravity	0.60

PIPE SIZE (in.)						
Nominal OD	$^1/_2$	$^3/_4$	1	$1^1/_4$	$1^1/_2$	2
Designation	SDR 9.33	SDR 11.0	SDR 11.00	SDR 10.00	SDR 11.00	SDR 11.00
Actual ID	0.660	0.860	1.077	1.328	1.554	1.943
Length (ft)	Capacity in Cubic Feet of Gas per Hour					
10	153	305	551	955	1,440	2,590
20	105	210	379	656	991	1,780
30	84	169	304	527	796	1,430
40	72	144	260	451	681	1,220
50	64	128	231	400	604	1,080
60	58	116	209	362	547	983
70	53	107	192	333	503	904
80	50	99	179	310	468	841
90	46	93	168	291	439	789
100	44	88	159	275	415	745
125	39	78	141	243	368	661
150	35	71	127	221	333	598
175	32	65	117	203	306	551
200	30	60	109	189	285	512
250	27	54	97	167	253	454
300	24	48	88	152	229	411
350	22	45	81	139	211	378
400	21	42	75	130	196	352
450	19	39	70	122	184	330
500	18	37	66	115	174	312

For SI: 1 inch = 25.4 mm, 1 foot = 304.8 mm, 1 pound per square inch = 6.895 kPa, 1-inch water column = 0.2488 kPa,
1 British thermal unit per hour = 0.2931 W, 1 cubic foot per hour = 0.0283 m³/h, 1 degree = 0.01745 rad.

Note: All table entries have been rounded to three significant digits.

**TABLE 402.4(19)
POLYETHYLENE PLASTIC PIPE**

Gas	Natural
Inlet Pressure	Less than 2 psi
Pressure Drop	0.5 in. w.c.
Specific Gravity	0.60

	PIPE SIZE (in.)					
Nominal OD	$^1/_2$	$^3/_4$	1	$1^1/_4$	$1^1/_2$	2
Designation	SDR 9.33	SDR 11.0	SDR 11.00	SDR 10.00	SDR 11.00	SDR 11.00
Actual ID	0.660	0.860	1.077	1.328	1.554	1.943
Length (ft)	Capacity in Cubic Feet of Gas per Hour					
10	201	403	726	1,260	1,900	3,410
20	138	277	499	865	1,310	2,350
30	111	222	401	695	1,050	1,880
40	95	190	343	594	898	1,610
50	84	169	304	527	796	1,430
60	76	153	276	477	721	1,300
70	70	140	254	439	663	1,190
80	65	131	236	409	617	1,110
90	61	123	221	383	579	1,040
100	58	116	209	362	547	983
125	51	103	185	321	485	871
150	46	93	168	291	439	789
175	43	86	154	268	404	726
200	40	80	144	249	376	675
250	35	71	127	221	333	598
300	32	64	115	200	302	542
350	29	59	106	184	278	499
400	27	55	99	171	258	464
450	26	51	93	160	242	435
500	24	48	88	152	229	411

For SI: 1 inch = 25.4 mm, 1 foot = 304.8 mm, 1 pound per square inch = 6.895 kPa, 1-inch water column = 0.2488 kPa,
1 British thermal unit per hour = 0.2931 W, 1 cubic foot per hour = 0.0283 m³/h, 1 degree = 0.01745 rad.

Note: All table entries have been rounded to three significant digits.

**TABLE 402.4(20)
POLYETHYLENE PLASTIC PIPE**

Gas	Natural
Inlet Pressure	2.0 psi
Pressure Drop	1.0 psi
Specific Gravity	0.60

	PIPE SIZE (in.)					
Nominal OD	$^1/_2$	$^3/_4$	1	$1^1/_4$	$1^1/_2$	2
Designation	SDR 9.33	SDR 11.0	SDR 11.00	SDR 10.00	SDR 11.00	SDR 11.00
Actual ID	0.660	0.860	1.077	1.328	1.554	1.943
Length (ft)	Capacity in Cubic Feet of Gas per Hour					
10	1,860	3,720	6,710	11,600	17,600	31,600
20	1,280	2,560	4,610	7,990	12,100	21,700
30	1,030	2,050	3,710	6,420	9,690	17,400
40	878	1,760	3,170	5,490	8,300	14,900
50	778	1,560	2,810	4,870	7,350	13,200
60	705	1,410	2,550	4,410	6,660	12,000
70	649	1,300	2,340	4,060	6,130	11,000
80	603	1,210	2,180	3,780	5,700	10,200
90	566	1,130	2,050	3,540	5,350	9,610
100	535	1,070	1,930	3,350	5,050	9,080
125	474	949	1,710	2,970	4,480	8,050
150	429	860	1,550	2,690	4,060	7,290
175	395	791	1,430	2,470	3,730	6,710
200	368	736	1,330	2,300	3,470	6,240
250	326	652	1,180	2,040	3,080	5,530
300	295	591	1,070	1,850	2,790	5,010
350	272	544	981	1,700	2,570	4,610
400	253	506	913	1,580	2,390	4,290
450	237	475	856	1,480	2,240	4,020
500	224	448	809	1,400	2,120	3,800
550	213	426	768	1,330	2,010	3,610
600	203	406	733	1,270	1,920	3,440
650	194	389	702	1,220	1,840	3,300
700	187	374	674	1,170	1,760	3,170
750	180	360	649	1,130	1,700	3,050
800	174	348	627	1,090	1,640	2,950
850	168	336	607	1,050	1,590	2,850
900	163	326	588	1,020	1,540	2,770
950	158	317	572	990	1,500	2,690
1,000	154	308	556	963	1,450	2,610
1,100	146	293	528	915	1,380	2,480
1,200	139	279	504	873	1,320	2,370
1,300	134	267	482	836	1,260	2,270
1,400	128	257	463	803	1,210	2,180
1,500	124	247	446	773	1,170	2,100
1,600	119	239	431	747	1,130	2,030
1,700	115	231	417	723	1,090	1,960
1,800	112	224	404	701	1,060	1,900
1,900	109	218	393	680	1,030	1,850
2,000	106	212	382	662	1,000	1,800

For SI: 1 inch = 25.4 mm, 1 foot = 304.8 mm, 1 pound per square inch = 6.895 kPa, 1-inch water column = 0.2488 kPa,
1 British thermal unit per hour = 0.2931 W, 1 cubic foot per hour = 0.0283 m³/h, 1 degree = 0.01745 rad.

Note: All table entries have been rounded to three significant digits.

TABLE 402.4(21)
POLYETHYLENE PLASTIC TUBING

Gas	Natural
Inlet Pressure	Less than 2.0 psi
Pressure Drop	0.3 in. w.c.
Specific Gravity	0.60

	PLASTIC TUBING SIZE (CTS) (in.)	
Nominal OD	$^1/_2$	$^3/_4$
Designation	SDR 7.00	SDR 11.00
Actual ID	0.445	0.927
Length (ft)	Capacity in Cubic Feet of Gas per Hour	
10	54	372
20	37	256
30	30	205
40	26	176
50	23	156
60	21	141
70	19	130
80	18	121
90	17	113
100	16	107
125	14	95
150	13	86
175	12	79
200	11	74
225	10	69
250	NA	65
275	NA	62
300	NA	59
350	NA	54
400	NA	51
450	NA	47
500	NA	45

For SI: 1 inch = 25.4 mm, 1 foot = 304.8 mm,
1 pound per square inch = 6.895 kPa, 1-inch water column = 0.2488 kPa,
1 British thermal unit per hour = 0.2931 W,
1 cubic foot per hour = 0.0283 m³/h, 1 degree = 0.01745 rad.
Notes:
1. NA means a flow of less than 10 cfh.
2. All table entries have been rounded to three significant digits.

TABLE 402.4(22)
POLYETHYLENE PLASTIC TUBING

Gas	Natural
Inlet Pressure	Less than 2.0 psi
Pressure Drop	0.5 in. w.c.
Specific Gravity	0.60

	PLASTIC TUBING SIZE (CTS) (in.)	
Nominal OD	$^1/_2$	$^3/_4$
Designation	SDR 7.00	SDR 11.00
Actual ID	0.445	0.927
Length (ft)	Capacity in Cubic Feet of Gas per Hour	
10	72	490
20	49	337
30	39	271
40	34	232
50	30	205
60	27	186
70	25	171
80	23	159
90	22	149
100	21	141
125	18	125
150	17	113
175	15	104
200	14	97
225	13	91
250	12	86
275	11	82
300	11	78
350	10	72
400	NA	67
450	NA	63
500	NA	59

For SI: 1 inch = 25.4 mm, 1 foot = 304.8 mm,
1 pound per square inch = 6.895 kPa, 1-inch water column = 0.2488 kPa,
1 British thermal unit per hour = 0.2931 W,
1 cubic foot per hour = 0.0283 m³/h, 1 degree = 0.01745 rad.
Notes:
1. NA means a flow of less than 10 cfh.
2. All table entries have been rounded to three significant digits.

TABLE 402.4(23)
SCHEDULE 40 METALLIC PIPE

Gas	Undiluted Propane
Inlet Pressure	10.0 psi
Pressure Drop	1.0 psi
Specific Gravity	1.50

SPECIAL USE	Pipe sizing between first stage (high-pressure regulator) and second stage (low-pressure regulator).								
	PIPE SIZE (in.)								
Nominal	$^1/_2$	$^3/_4$	1	$1^1/_4$	$1^1/_2$	2	$2^1/_2$	3	4
Actual ID	0.622	0.824	1.049	1.380	1.610	2.067	2.469	3.068	4.026
Length (ft)	Capacity in Thousands of Btu per Hour								
10	3,320	6,950	13,100	26,900	40,300	77,600	124,000	219,000	446,000
20	2,280	4,780	9,000	18,500	27,700	53,300	85,000	150,000	306,000
30	1,830	3,840	7,220	14,800	22,200	42,800	68,200	121,000	246,000
40	1,570	3,280	6,180	12,700	19,000	36,600	58,400	103,000	211,000
50	1,390	2,910	5,480	11,300	16,900	32,500	51,700	91,500	187,000
60	1,260	2,640	4,970	10,200	15,300	29,400	46,900	82,900	169,000
70	1,160	2,430	4,570	9,380	14,100	27,100	43,100	76,300	156,000
80	1,080	2,260	4,250	8,730	13,100	25,200	40,100	70,900	145,000
90	1,010	2,120	3,990	8,190	12,300	23,600	37,700	66,600	136,000
100	956	2,000	3,770	7,730	11,600	22,300	35,600	62,900	128,000
125	848	1,770	3,340	6,850	10,300	19,800	31,500	55,700	114,000
150	768	1,610	3,020	6,210	9,300	17,900	28,600	50,500	103,000
175	706	1,480	2,780	5,710	8,560	16,500	26,300	46,500	94,700
200	657	1,370	2,590	5,320	7,960	15,300	24,400	43,200	88,100
250	582	1,220	2,290	4,710	7,060	13,600	21,700	38,300	78,100
300	528	1,100	2,080	4,270	6,400	12,300	19,600	34,700	70,800
350	486	1,020	1,910	3,930	5,880	11,300	18,100	31,900	65,100
400	452	945	1,780	3,650	5,470	10,500	16,800	29,700	60,600
450	424	886	1,670	3,430	5,140	9,890	15,800	27,900	56,800
500	400	837	1,580	3,240	4,850	9,340	14,900	26,300	53,700
550	380	795	1,500	3,070	4,610	8,870	14,100	25,000	51,000
600	363	759	1,430	2,930	4,400	8,460	13,500	23,900	48,600
650	347	726	1,370	2,810	4,210	8,110	12,900	22,800	46,600
700	334	698	1,310	2,700	4,040	7,790	12,400	21,900	44,800
750	321	672	1,270	2,600	3,900	7,500	12,000	21,100	43,100
800	310	649	1,220	2,510	3,760	7,240	11,500	20,400	41,600
850	300	628	1,180	2,430	3,640	7,010	11,200	19,800	40,300
900	291	609	1,150	2,360	3,530	6,800	10,800	19,200	39,100
950	283	592	1,110	2,290	3,430	6,600	10,500	18,600	37,900
1,000	275	575	1,080	2,230	3,330	6,420	10,200	18,100	36,900
1,100	261	546	1,030	2,110	3,170	6,100	9,720	17,200	35,000
1,200	249	521	982	2,020	3,020	5,820	9,270	16,400	33,400
1,300	239	499	940	1,930	2,890	5,570	8,880	15,700	32,000
1,400	229	480	903	1,850	2,780	5,350	8,530	15,100	30,800
1,500	221	462	870	1,790	2,680	5,160	8,220	14,500	29,600
1,600	213	446	840	1,730	2,590	4,980	7,940	14,000	28,600
1,700	206	432	813	1,670	2,500	4,820	7,680	13,600	27,700
1,800	200	419	789	1,620	2,430	4,670	7,450	13,200	26,900
1,900	194	407	766	1,570	2,360	4,540	7,230	12,800	26,100
2,000	189	395	745	1,530	2,290	4,410	7,030	12,400	25,400

For SI: 1 inch = 25.4 mm, 1 foot = 304.8 mm, 1 pound per square inch = 6.895 kPa, 1-inch water column = 0.2488 kPa,
1 British thermal unit per hour = 0.2931 W, 1 cubic foot per hour = 0.0283 m^3/h, 1 degree = 0.01745 rad.

Note: All table entries have been rounded to three significant digits.

	Gas	Undiluted Propane
	Inlet Pressure	10.0 psi
	Pressure Drop	3.0 psi
	Specific Gravity	1.50

TABLE 402.4(24)
SCHEDULE 40 METALLIC PIPE

SPECIAL USE	Pipe sizing between first stage (high-pressure regulator) and second stage (low-pressure regulator).								
PIPE SIZE (in)									
Nominal	$^1/_2$	$^3/_4$	1	$1^1/_4$	$1^1/_2$	2	$2^1/_2$	3	4
Actual ID	0.622	0.824	1.049	1.380	1.610	2.067	2.469	3.068	4.026
Length (ft)	Capacity in Thousands of Btu per Hour								
10	5,890	12,300	23,200	47,600	71,300	137,000	219,000	387,000	789,000
20	4,050	8,460	15,900	32,700	49,000	94,400	150,000	266,000	543,000
30	3,250	6,790	12,800	26,300	39,400	75,800	121,000	214,000	436,000
40	2,780	5,810	11,000	22,500	33,700	64,900	103,000	183,000	373,000
50	2,460	5,150	9,710	19,900	29,900	57,500	91,600	162,000	330,000
60	2,230	4,670	8,790	18,100	27,100	52,100	83,000	147,000	299,000
70	2,050	4,300	8,090	16,600	24,900	47,900	76,400	135,000	275,000
80	1,910	4,000	7,530	15,500	23,200	44,600	71,100	126,000	256,000
90	1,790	3,750	7,060	14,500	21,700	41,800	66,700	118,000	240,000
100	1,690	3,540	6,670	13,700	20,500	39,500	63,000	111,000	227,000
125	1,500	3,140	5,910	12,100	18,200	35,000	55,800	98,700	201,000
150	1,360	2,840	5,360	11,000	16,500	31,700	50,600	89,400	182,000
175	1,250	2,620	4,930	10,100	15,200	29,200	46,500	82,300	167,800
200	1,160	2,430	4,580	9,410	14,100	27,200	43,300	76,500	156,100
250	1,030	2,160	4,060	8,340	12,500	24,100	38,400	67,800	138,400
300	935	1,950	3,680	7,560	11,300	21,800	34,800	61,500	125,400
350	860	1,800	3,390	6,950	10,400	20,100	32,000	56,500	115,300
400	800	1,670	3,150	6,470	9,690	18,700	29,800	52,600	107,300
450	751	1,570	2,960	6,070	9,090	17,500	27,900	49,400	100,700
500	709	1,480	2,790	5,730	8,590	16,500	26,400	46,600	95,100
550	673	1,410	2,650	5,450	8,160	15,700	25,000	44,300	90,300
600	642	1,340	2,530	5,200	7,780	15,000	23,900	42,200	86,200
650	615	1,290	2,420	4,980	7,450	14,400	22,900	40,500	82,500
700	591	1,240	2,330	4,780	7,160	13,800	22,000	38,900	79,300
750	569	1,190	2,240	4,600	6,900	13,300	21,200	37,400	76,400
800	550	1,150	2,170	4,450	6,660	12,800	20,500	36,200	73,700
850	532	1,110	2,100	4,300	6,450	12,400	19,800	35,000	71,400
900	516	1,080	2,030	4,170	6,250	12,000	19,200	33,900	69,200
950	501	1,050	1,970	4,050	6,070	11,700	18,600	32,900	67,200
1,000	487	1,020	1,920	3,940	5,900	11,400	18,100	32,000	65,400
1,100	463	968	1,820	3,740	5,610	10,800	17,200	30,400	62,100
1,200	442	923	1,740	3,570	5,350	10,300	16,400	29,000	59,200
1,300	423	884	1,670	3,420	5,120	9,870	15,700	27,800	56,700
1,400	406	849	1,600	3,280	4,920	9,480	15,100	26,700	54,500
1,500	391	818	1,540	3,160	4,740	9,130	14,600	25,700	52,500
1,600	378	790	1,490	3,060	4,580	8,820	14,100	24,800	50,700
1,700	366	765	1,440	2,960	4,430	8,530	13,600	24,000	49,000
1,800	355	741	1,400	2,870	4,300	8,270	13,200	23,300	47,600
1,900	344	720	1,360	2,780	4,170	8,040	12,800	22,600	46,200
2,000	335	700	1,320	2,710	4,060	7,820	12,500	22,000	44,900

For SI: 1 inch = 25.4 mm, 1 foot = 304.8 mm, 1 pound per square inch = 6.895 kPa, 1-inch water column = 0.2488 kPa,
1 British thermal unit per hour = 0.2931 W, 1 cubic foot per hour = 0.0283 m³/h, 1 degree = 0.01745 rad.
Note: All table entries have been rounded to three significant digits.

TABLE 402.4(25)
SCHEDULE 40 METALLIC PIPE

Gas	Undiluted Propane
Inlet Pressure	2.0 psi
Pressure Drop	1.0 psi
Specific Gravity	1.50

PIPE SIZE (in.)									
Nominal	1/2	3/4	1	1 1/4	1 1/2	2	2 1/2	3	4
Actual ID	0.622	0.824	1.049	1.380	1.610	2.067	2.469	3.068	4.026
Length (ft)	Capacity in Thousands of Btu per Hour								
10	2,680	5,590	10,500	21,600	32,400	62,400	99,500	176,000	359,000
20	1,840	3,850	7,240	14,900	22,300	42,900	68,400	121,000	247,000
30	1,480	3,090	5,820	11,900	17,900	34,500	54,900	97,100	198,000
40	1,260	2,640	4,980	10,200	15,300	29,500	47,000	83,100	170,000
50	1,120	2,340	4,410	9,060	13,600	26,100	41,700	73,700	150,000
60	1,010	2,120	4,000	8,210	12,300	23,700	37,700	66,700	136,000
70	934	1,950	3,680	7,550	11,300	21,800	34,700	61,400	125,000
80	869	1,820	3,420	7,020	10,500	20,300	32,300	57,100	116,000
90	815	1,700	3,210	6,590	9,880	19,000	30,300	53,600	109,000
100	770	1,610	3,030	6,230	9,330	18,000	28,600	50,600	103,000
125	682	1,430	2,690	5,520	8,270	15,900	25,400	44,900	91,500
150	618	1,290	2,440	5,000	7,490	14,400	23,000	40,700	82,900
175	569	1,190	2,240	4,600	6,890	13,300	21,200	37,400	76,300
200	529	1,110	2,080	4,280	6,410	12,300	19,700	34,800	71,000
250	469	981	1,850	3,790	5,680	10,900	17,400	30,800	62,900
300	425	889	1,670	3,440	5,150	9,920	15,800	27,900	57,000
350	391	817	1,540	3,160	4,740	9,120	14,500	25,700	52,400
400	364	760	1,430	2,940	4,410	8,490	13,500	23,900	48,800
450	341	714	1,340	2,760	4,130	7,960	12,700	22,400	45,800
500	322	674	1,270	2,610	3,910	7,520	12,000	21,200	43,200
550	306	640	1,210	2,480	3,710	7,140	11,400	20,100	41,100
600	292	611	1,150	2,360	3,540	6,820	10,900	19,200	39,200
650	280	585	1,100	2,260	3,390	6,530	10,400	18,400	37,500
700	269	562	1,060	2,170	3,260	6,270	9,990	17,700	36,000
750	259	541	1,020	2,090	3,140	6,040	9,630	17,000	34,700
800	250	523	985	2,020	3,030	5,830	9,300	16,400	33,500
850	242	506	953	1,960	2,930	5,640	9,000	15,900	32,400
900	235	490	924	1,900	2,840	5,470	8,720	15,400	31,500
950	228	476	897	1,840	2,760	5,310	8,470	15,000	30,500
1,000	222	463	873	1,790	2,680	5,170	8,240	14,600	29,700
1,100	210	440	829	1,700	2,550	4,910	7,830	13,800	28,200
1,200	201	420	791	1,620	2,430	4,680	7,470	13,200	26,900
1,300	192	402	757	1,550	2,330	4,490	7,150	12,600	25,800
1,400	185	386	727	1,490	2,240	4,310	6,870	12,100	24,800
1,500	178	372	701	1,440	2,160	4,150	6,620	11,700	23,900
1,600	172	359	677	1,390	2,080	4,010	6,390	11,300	23,000
1,700	166	348	655	1,340	2,010	3,880	6,180	10,900	22,300
1,800	161	337	635	1,300	1,950	3,760	6,000	10,600	21,600
1,900	157	327	617	1,270	1,900	3,650	5,820	10,300	21,000
2,000	152	318	600	1,230	1,840	3,550	5,660	10,000	20,400

For SI: 1 inch = 25.4 mm, 1 foot = 304.8 mm, 1 pound per square inch = 6.895 kPa, 1-inch water column = 0.2488 kPa,
1 British thermal unit per hour = 0.2931 W, 1 cubic foot per hour = 0.0283 m³/h, 1 degree = 0.01745 rad.

Note: All table entries have been rounded to three significant digits.

		Gas	Undiluted Propane
		Inlet Pressure	11.0 in. w.c.
		Pressure Drop	0.5 in. w.c.
		Specific Gravity	1.50

TABLE 402.4(26)
SCHEDULE 40 METALLIC PIPE

SPECIAL USE	Pipe sizing between first stage (high-pressure regulator) and second stage (low-pressure regulator).								
PIPE SIZE (in.)									
Nominal	$^1/_2$	$^3/_4$	1	$1^1/_4$	$1^1/_2$	2	$2^1/_2$	3	4
Actual ID	0.622	0.824	1.049	1.380	1.610	2.067	2.469	3.068	4.026
Length (ft)	**Capacity in Thousands of Btu per Hour**								
10	291	608	1,150	2,350	3,520	6,790	10,800	19,100	39,000
20	200	418	787	1,620	2,420	4,660	7,430	13,100	26,800
30	160	336	632	1,300	1,940	3,750	5,970	10,600	21,500
40	137	287	541	1,110	1,660	3,210	5,110	9,030	18,400
50	122	255	480	985	1,480	2,840	4,530	8,000	16,300
60	110	231	434	892	1,340	2,570	4,100	7,250	14,800
80	101	212	400	821	1,230	2,370	3,770	6,670	13,600
100	94	197	372	763	1,140	2,200	3,510	6,210	12,700
125	89	185	349	716	1,070	2,070	3,290	5,820	11,900
150	84	175	330	677	1,010	1,950	3,110	5,500	11,200
175	74	155	292	600	899	1,730	2,760	4,880	9,950
200	67	140	265	543	814	1,570	2,500	4,420	9,010
250	62	129	243	500	749	1,440	2,300	4,060	8,290
300	58	120	227	465	697	1,340	2,140	3,780	7,710
350	51	107	201	412	618	1,190	1,900	3,350	6,840
400	46	97	182	373	560	1,080	1,720	3,040	6,190
450	42	89	167	344	515	991	1,580	2,790	5,700
500	40	83	156	320	479	922	1,470	2,600	5,300
550	37	78	146	300	449	865	1,380	2,440	4,970
600	35	73	138	283	424	817	1,300	2,300	4,700
650	33	70	131	269	403	776	1,240	2,190	4,460
700	32	66	125	257	385	741	1,180	2,090	4,260
750	30	64	120	246	368	709	1,130	2,000	4,080
800	29	61	115	236	354	681	1,090	1,920	3,920
850	28	59	111	227	341	656	1,050	1,850	3,770
900	27	57	107	220	329	634	1,010	1,790	3,640
950	26	55	104	213	319	613	978	1,730	3,530
1,000	25	53	100	206	309	595	948	1,680	3,420
1,100	25	52	97	200	300	578	921	1,630	3,320
1,200	24	50	95	195	292	562	895	1,580	3,230
1,300	23	48	90	185	277	534	850	1,500	3,070
1,400	22	46	86	176	264	509	811	1,430	2,930
1,500	21	44	82	169	253	487	777	1,370	2,800
1,600	20	42	79	162	243	468	746	1,320	2,690
1,700	19	40	76	156	234	451	719	1,270	2,590
1,800	19	39	74	151	226	436	694	1,230	2,500
1,900	18	38	71	146	219	422	672	1,190	2,420
2,000	18	37	69	142	212	409	652	1,150	2,350

For SI: 1 inch = 25.4 mm, 1 foot = 304.8 mm, 1 pound per square inch = 6.895 kPa, 1-inch water column = 0.2488 kPa,
1 British thermal unit per hour = 0.2931 W, 1 cubic foot per hour = 0.0283 m³/h, 1 degree = 0.01745 rad.

Note: All table entries have been rounded to three significant digits.

TABLE 402.4(27)
SEMIRIGID COPPER TUBING

Gas	Undiluted Propane
Inlet Pressure	10.0 psi
Pressure Drop	1.0 psi
Specific Gravity	1.50

SPECIAL USE		Sizing between first stage (high-pressure regulator) and second stage (low-pressure regulator).								
TUBE SIZE (in.)										
Nominal	**K & L**	$^1/_4$	$^3/_8$	$^1/_2$	$^5/_8$	$^3/_4$	1	$1^1/_4$	$1^1/_2$	2
	ACR	$^3/_8$	$^1/_2$	$^5/_8$	$^3/_4$	$^7/_8$	$1^1/_8$	$1^3/_8$	—	—
Outside		0.375	0.500	0.625	0.750	0.875	1.125	1.375	1.625	2.125
Inside		0.305	0.402	0.527	0.652	0.745	0.995	1.245	1.481	1.959
Length (ft)		**Capacity in Thousands of Btu per Hour**								
10		513	1,060	2,150	3,760	5,330	11,400	20,500	32,300	67,400
20		352	727	1,480	2,580	3,670	7,830	14,100	22,200	46,300
30		283	584	1,190	2,080	2,940	6,290	11,300	17,900	37,200
40		242	500	1,020	1,780	2,520	5,380	9,690	15,300	31,800
50		215	443	901	1,570	2,230	4,770	8,590	13,500	28,200
60		194	401	816	1,430	2,020	4,320	7,780	12,300	25,600
70		179	369	751	1,310	1,860	3,980	7,160	11,300	23,500
80		166	343	699	1,220	1,730	3,700	6,660	10,500	21,900
90		156	322	655	1,150	1,630	3,470	6,250	9,850	20,500
100		147	304	619	1,080	1,540	3,280	5,900	9,310	19,400
125		131	270	549	959	1,360	2,910	5,230	8,250	17,200
150		118	244	497	869	1,230	2,630	4,740	7,470	15,600
175		109	225	457	799	1,130	2,420	4,360	6,880	14,300
200		101	209	426	744	1,060	2,250	4,060	6,400	13,300
250		90	185	377	659	935	2,000	3,600	5,670	11,800
300		81	168	342	597	847	1,810	3,260	5,140	10,700
350		75	155	314	549	779	1,660	3,000	4,730	9,840
400		70	144	292	511	725	1,550	2,790	4,400	9,160
450		65	135	274	480	680	1,450	2,620	4,130	8,590
500		62	127	259	453	643	1,370	2,470	3,900	8,120
550		59	121	246	430	610	1,300	2,350	3,700	7,710
600		56	115	235	410	582	1,240	2,240	3,530	7,350
650		54	111	225	393	558	1,190	2,140	3,380	7,040
700		51	106	216	378	536	1,140	2,060	3,250	6,770
750		50	102	208	364	516	1,100	1,980	3,130	6,520
800		48	99	201	351	498	1,060	1,920	3,020	6,290
850		46	96	195	340	482	1,030	1,850	2,920	6,090
900		45	93	189	330	468	1,000	1,800	2,840	5,910
950		44	90	183	320	454	970	1,750	2,750	5,730
1,000		42	88	178	311	442	944	1,700	2,680	5,580
1,100		40	83	169	296	420	896	1,610	2,540	5,300
1,200		38	79	161	282	400	855	1,540	2,430	5,050
1,300		37	76	155	270	383	819	1,470	2,320	4,840
1,400		35	73	148	260	368	787	1,420	2,230	4,650
1,500		34	70	143	250	355	758	1,360	2,150	4,480
1,600		33	68	138	241	343	732	1,320	2,080	4,330
1,700		32	66	134	234	331	708	1,270	2,010	4,190
1,800		31	64	130	227	321	687	1,240	1,950	4,060
1,900		30	62	126	220	312	667	1,200	1,890	3,940
2,000		29	60	122	214	304	648	1,170	1,840	3,830

For SI: 1 inch = 25.4 mm, 1 foot = 304.8 mm, 1 pound per square inch = 6.895 kPa, 1-inch water column = 0.2488 kPa,
1 British thermal unit per hour = 0.2931 W, 1 cubic foot per hour = 0.0283 m³/h, 1 degree = 0.01745 rad.

Notes:
1. Table capacities are based on Type K copper tubing inside diameter (shown), which has the smallest inside diameter of the copper tubing products.
2. All table entries have been rounded to three significant digits.

Gas	Undiluted Propane
Inlet Pressure	11.0 in. w.c.
Pressure Drop	0.5 in. w.c.
Specific Gravity	1.50

TABLE 402.4(28)
SEMIRIGID COPPER TUBING

SPECIAL USE		Sizing between first stage (high-pressure regulator) and second stage (low-pressure regulator)								
		TUBE SIZE (in.)								
Nominal	K & L	$1/4$	$3/8$	$1/2$	$5/8$	$3/4$	1	$1^1/4$	$1^1/2$	2
	ACR	$3/8$	$1/2$	$5/8$	$3/4$	$7/8$	$1^1/8$	$1^3/8$	—	—
Outside		0.375	0.500	0.625	0.750	0.875	1.125	1.375	1.625	2.125
Inside		0.305	0.402	0.527	0.652	0.745	0.995	1.245	1.481	1.959
Length (ft)		Capacity in Thousands of Btu per Hour								
10		45	93	188	329	467	997	1,800	2,830	5,890
20		31	64	129	226	321	685	1,230	1,950	4,050
30		25	51	104	182	258	550	991	1,560	3,250
40		21	44	89	155	220	471	848	1,340	2,780
50		19	39	79	138	195	417	752	1,180	2,470
60		17	35	71	125	177	378	681	1,070	2,240
70		16	32	66	115	163	348	626	988	2,060
80		15	30	61	107	152	324	583	919	1,910
90		14	28	57	100	142	304	547	862	1,800
100		13	27	54	95	134	287	517	814	1,700
125		11	24	48	84	119	254	458	722	1,500
150		10	21	44	76	108	230	415	654	1,360
175		NA	20	40	70	99	212	382	602	1,250
200		NA	18	37	65	92	197	355	560	1,170
250		NA	16	33	58	82	175	315	496	1,030
300		NA	15	30	52	74	158	285	449	936
350		NA	14	28	48	68	146	262	414	861
400		NA	13	26	45	63	136	244	385	801
450		NA	12	24	42	60	127	229	361	752
500		NA	11	23	40	56	120	216	341	710
550		NA	11	22	38	53	114	205	324	674
600		NA	10	21	36	51	109	196	309	643
650		NA	NA	20	34	49	104	188	296	616
700		NA	NA	19	33	47	100	180	284	592
750		NA	NA	18	32	45	96	174	274	570
800		NA	NA	18	31	44	93	168	264	551
850		NA	NA	17	30	42	90	162	256	533
900		NA	NA	17	29	41	87	157	248	517
950		NA	NA	16	28	40	85	153	241	502
1,000		NA	NA	16	27	39	83	149	234	488
1,100		NA	NA	15	26	37	78	141	223	464
1,200		NA	NA	14	25	35	75	135	212	442
1,300		NA	NA	14	24	34	72	129	203	423
1,400		NA	NA	13	23	32	69	124	195	407
1,500		NA	NA	13	22	31	66	119	188	392
1,600		NA	NA	12	21	30	64	115	182	378
1,700		NA	NA	12	20	29	62	112	176	366
1,800		NA	NA	11	20	28	60	108	170	355
1,900		NA	NA	11	19	27	58	105	166	345
2,000		NA	NA	11	19	27	57	102	161	335

For SI: 1 inch = 25.4 mm, 1 foot = 304.8 mm, 1 pound per square inch = 6.895 kPa, 1-inch water column = 0.2488 kPa,
1 British thermal unit per hour = 0.2931 W, 1 cubic foot per hour = 0.0283 m³/h, 1 degree = 0.01745 rad.

Notes:
1. Table capacities are based on Type K copper tubing inside diameter (shown), which has the smallest inside diameter of the copper tubing products.
2. NA means a flow of less than 10,000 Btu/hr.
3. All table entries have been rounded to three significant digits.

	Gas	Undiluted Propane
	Inlet Pressure	2.0 psi
	Pressure Drop	1.0 psi
	Specific Gravity	1.50

TABLE 402.4(29)
SEMIRIGID COPPER TUBING

TUBE SIZE (in.)										
Nominal	K & L	$1/4$	$3/8$	$1/2$	$5/8$	$3/4$	1	$1^1/4$	$1^1/2$	2
	ACR	$3/8$	$1/2$	$5/8$	$3/4$	$7/8$	$1^1/8$	$1^3/8$	—	—
Outside		0.375	0.500	0.625	0.750	0.875	1.125	1.375	1.625	2.125
Inside		0.305	0.402	0.527	0.652	0.745	0.995	1.245	1.481	1.959
Length (ft)		Capacity in Thousands of Btu per Hour								
10		413	852	1,730	3,030	4,300	9,170	16,500	26,000	54,200
20		284	585	1,190	2,080	2,950	6,310	11,400	17,900	37,300
30		228	470	956	1,670	2,370	5,060	9,120	14,400	29,900
40		195	402	818	1,430	2,030	4,330	7,800	12,300	25,600
50		173	356	725	1,270	1,800	3,840	6,920	10,900	22,700
60		157	323	657	1,150	1,630	3,480	6,270	9,880	20,600
70		144	297	605	1,060	1,500	3,200	5,760	9,090	18,900
80		134	276	562	983	1,390	2,980	5,360	8,450	17,600
90		126	259	528	922	1,310	2,790	5,030	7,930	16,500
100		119	245	498	871	1,240	2,640	4,750	7,490	15,600
125		105	217	442	772	1,100	2,340	4,210	6,640	13,800
150		95	197	400	700	992	2,120	3,820	6,020	12,500
175		88	181	368	644	913	1,950	3,510	5,540	11,500
200		82	168	343	599	849	1,810	3,270	5,150	10,700
250		72	149	304	531	753	1,610	2,900	4,560	9,510
300		66	135	275	481	682	1,460	2,620	4,140	8,610
350		60	124	253	442	628	1,340	2,410	3,800	7,920
400		56	116	235	411	584	1,250	2,250	3,540	7,370
450		53	109	221	386	548	1,170	2,110	3,320	6,920
500		50	103	209	365	517	1,110	1,990	3,140	6,530
550		47	97	198	346	491	1,050	1,890	2,980	6,210
600		45	93	189	330	469	1,000	1,800	2,840	5,920
650		43	89	181	316	449	959	1,730	2,720	5,670
700		41	86	174	304	431	921	1,660	2,620	5,450
750		40	82	168	293	415	888	1,600	2,520	5,250
800		39	80	162	283	401	857	1,540	2,430	5,070
850		37	77	157	274	388	829	1,490	2,350	4,900
900		36	75	152	265	376	804	1,450	2,280	4,750
950		35	72	147	258	366	781	1,410	2,220	4,620
1,000		34	71	143	251	356	760	1,370	2,160	4,490
1,100		32	67	136	238	338	721	1,300	2,050	4,270
1,200		31	64	130	227	322	688	1,240	1,950	4,070
1,300		30	61	124	217	309	659	1,190	1,870	3,900
1,400		28	59	120	209	296	633	1,140	1,800	3,740
1,500		27	57	115	201	286	610	1,100	1,730	3,610
1,600		26	55	111	194	276	589	1,060	1,670	3,480
1,700		26	53	108	188	267	570	1,030	1,620	3,370
1,800		25	51	104	182	259	553	1,000	1,570	3,270
1,900		24	50	101	177	251	537	966	1,520	3,170
2,000		23	48	99	172	244	522	940	1,480	3,090

For SI: 1 inch = 25.4 mm, 1 foot = 304.8 mm, 1 pound per square inch = 6.895 kPa, 1-inch water column = 0.2488 kPa,
1 British thermal unit per hour = 0.2931 W, 1 cubic foot per hour = 0.0283 m^3/h, 1 degree = 0.01745 rad.

Notes:
1. Table capacities are based on Type K copper tubing inside diameter (shown), which has the smallest inside diameter of the copper tubing products.
2. All table entries have been rounded to three significant digits.

TABLE 402.4(30)
CORRUGATED STAINLESS STEEL TUBING (CSST)

Gas	Undiluted Propane
Inlet Pressure	11.0 in. w.c.
Pressure Drop	0.5 in. w.c.
Specific Gravity	1.50

Flow Designation	TUBE SIZE (EHD)												
	13	15	18	19	23	25	30	31	37	46	48	60	62
Length (ft)	Capacity in Thousands of Btu per Hour												
5	72	99	181	211	355	426	744	863	1,420	2,830	3,270	5,780	6,550
10	50	69	129	150	254	303	521	605	971	1,990	2,320	4,110	4,640
15	39	55	104	121	208	248	422	490	775	1,620	1,900	3,370	3,790
20	34	49	91	106	183	216	365	425	661	1,400	1,650	2,930	3,290
25	30	42	82	94	164	192	325	379	583	1,250	1,480	2,630	2,940
30	28	39	74	87	151	177	297	344	528	1,140	1,350	2,400	2,680
40	23	33	64	74	131	153	256	297	449	988	1,170	2,090	2,330
50	20	30	58	66	118	137	227	265	397	884	1,050	1,870	2,080
60	19	26	53	60	107	126	207	241	359	805	961	1,710	1,900
70	17	25	49	57	99	117	191	222	330	745	890	1,590	1,760
80	15	23	45	52	94	109	178	208	307	696	833	1,490	1,650
90	15	22	44	50	90	102	169	197	286	656	787	1,400	1,550
100	14	20	41	47	85	98	159	186	270	621	746	1,330	1,480
150	11	15	31	36	66	75	123	143	217	506	611	1,090	1,210
200	9	14	28	33	60	69	112	129	183	438	531	948	1,050
250	8	12	25	30	53	61	99	117	163	390	476	850	934
300	8	11	23	26	50	57	90	107	147	357	434	777	854

For SI: 1 inch = 25.4 mm, 1 foot = 304.8 mm, 1 pound per square inch = 6.895 kPa, 1-inch water column = 0.2488 kPa, 1 British thermal unit per hour = 0.2931 W, 1 cubic foot per hour = 0.0283 m³/h, 1 degree = 0.01745 rad.

Notes:

1. Table includes losses for four 90-degree bends and two end fittings. Tubing runs with larger numbers of bends and/or fittings shall be increased by an equivalent length of tubing to the following equation: $L = 1.3n$ where L is additional length (feet) of tubing and n is the number of additional fittings and/or bends.

2. EHD—Equivalent Hydraulic Diameter, which is a measure of the relative hydraulic efficiency between different tubing sizes. The greater the value of EHD, the greater the gas capacity of the tubing.

3. All table entries have been rounded to three significant digits.

TABLE 402.4(31)
CORRUGATED STAINLESS STEEL TUBING (CSST)

Gas	Undiluted Propane
Inlet Pressure	2.0 psi
Pressure Drop	1.0 psi
Specific Gravity	1.50

TUBE SIZE (EHD)													
Flow Designation	13	15	18	19	23	25	30	31	37	46	48	60	62
Length (ft)	Capacity in Thousands of Btu per Hour												
10	426	558	927	1,110	1,740	2,170	4,100	4,720	7,130	15,200	16,800	29,400	34,200
25	262	347	591	701	1,120	1,380	2,560	2,950	4,560	9,550	10,700	18,800	21,700
30	238	316	540	640	1,030	1,270	2,330	2,690	4,180	8,710	9,790	17,200	19,800
40	203	271	469	554	896	1,100	2,010	2,320	3,630	7,530	8,500	14,900	17,200
50	181	243	420	496	806	986	1,790	2,070	3,260	6,730	7,610	13,400	15,400
75	147	196	344	406	663	809	1,460	1,690	2,680	5,480	6,230	11,000	12,600
80	140	189	333	393	643	768	1,410	1,630	2,590	5,300	6,040	10,600	12,200
100	124	169	298	350	578	703	1,260	1,450	2,330	4,740	5,410	9,530	10,900
150	101	137	245	287	477	575	1,020	1,180	1,910	3,860	4,430	7,810	8,890
200	86	118	213	248	415	501	880	1,020	1,660	3,340	3,840	6,780	7,710
250	77	105	191	222	373	448	785	910	1,490	2,980	3,440	6,080	6,900
300	69	96	173	203	343	411	716	829	1,360	2,720	3,150	5,560	6,300
400	60	82	151	175	298	355	616	716	1,160	2,350	2,730	4,830	5,460
500	53	72	135	158	268	319	550	638	1,030	2,100	2,450	4,330	4,880

For SI: 1 inch = 25.4 mm, 1 foot = 304.8 mm, 1 pound per square inch = 6.895kPa, 1-inch water column = 0.2488 kPa, 1 British thermal unit per hour = 0.2931 W, 1 cubic foot per hour = 0.0283 m³/h, 1 degree = 0.01745 rad.

Notes:

1. Table does not include effect of pressure drop across the line regulator. Where regulator loss exceeds $^1/_2$ psi (based on 13 in. w.c. outlet pressure), DO NOT USE THIS TABLE. Consult with the regulator manufacturer for pressure drops and capacity factors. Pressure drops across a regulator may vary with flow rate.
2. CAUTION: Capacities shown in the table might exceed maximum capacity for a selected regulator. Consult with the regulator or tubing manufacturer for guidance.
3. Table includes losses for four 90-degree bends and two end fittings. Tubing runs with larger numbers of bends and/or fittings shall be increased by an equivalent length of tubing to the following equation: $L = 1.3n$ where L is additional length (feet) of tubing and n is the number of additional fittings and/or bends.
4. EHD—Equivalent Hydraulic Diameter, which is a measure of the relative hydraulic efficiency between different tubing sizes. The greater the value of EHD, the greater the gas capacity of the tubing.
5. All table entries have been rounded to three significant digits.

TABLE 402.4(32)
CORRUGATED STAINLESS STEEL TUBING (CSST)

Gas	Undiluted Propane
Inlet Pressure	5.0 psi
Pressure Drop	3.5 psi
Specific Gravity	1.50

TUBE SIZE (EHD)													
Flow Designation	13	15	18	19	23	25	30	31	37	46	48	60	62
Length (ft)	Capacity in Thousands of Btu per Hour												
10	826	1,070	1,710	2,060	3,150	4,000	7,830	8,950	13,100	28,600	31,200	54,400	63,800
25	509	664	1,090	1,310	2,040	2,550	4,860	5,600	8,400	18,000	19,900	34,700	40,400
30	461	603	999	1,190	1,870	2,340	4,430	5,100	7,680	16,400	18,200	31,700	36,900
40	396	520	867	1,030	1,630	2,030	3,820	4,400	6,680	14,200	15,800	27,600	32,000
50	352	463	777	926	1,460	1,820	3,410	3,930	5,990	12,700	14,100	24,700	28,600
75	284	376	637	757	1,210	1,490	2,770	3,190	4,920	10,300	11,600	20,300	23,400
80	275	363	618	731	1,170	1,450	2,680	3,090	4,770	9,990	11,200	19,600	22,700
100	243	324	553	656	1,050	1,300	2,390	2,760	4,280	8,930	10,000	17,600	20,300
150	196	262	453	535	866	1,060	1,940	2,240	3,510	7,270	8,210	14,400	16,600
200	169	226	393	464	755	923	1,680	1,930	3,050	6,290	7,130	12,500	14,400
250	150	202	352	415	679	828	1,490	1,730	2,740	5,620	6,390	11,200	12,900
300	136	183	322	379	622	757	1,360	1,570	2,510	5,120	5,840	10,300	11,700
400	117	158	279	328	542	657	1,170	1,360	2,180	4,430	5,070	8,920	10,200
500	104	140	251	294	488	589	1,050	1,210	1,950	3,960	4,540	8,000	9,110

For SI: 1 inch = 25.4 mm, 1 foot = 304.8 mm, 1 pound per square inch = 6.895 kPa, 1-inch water column = 0.2488 kPa, 1 British thermal unit per hour = 0.2931 W, 1 cubic foot per hour = 0.0283 m^3/h, 1 degree = 0.01745 rad.

Notes:

1. Table does not include effect of pressure drop across line regulator. Where regulator loss exceeds 1 psi, DO NOT USE THIS TABLE. Consult with the regulator manufacturer for pressure drops and capacity factors. Pressure drop across regulator may vary with the flow rate.

2. CAUTION: Capacities shown in the table might exceed maximum capacity of selected regulator. Consult with the tubing manufacturer for guidance.

3. Table includes losses for four 90-degree bends and two end fittings. Tubing runs with larger numbers of bends and/or fittings shall be increased by an equivalent length of tubing to the following equation: $L = 1.3n$ where L is additional length (feet) of tubing and n is the number of additional fittings and/or bends.

4. EHD— Equivalent Hydraulic Diameter, which is a measure of the relative hydraulic efficiency between different tubing sizes. The greater the value of EHD, the greater the gas capacity of the tubing.

5. All table entries have been rounded to three significant digits.

TABLE 402.4(33)
POLYETHYLENE PLASTIC PIPE

Gas	Undiluted Propane
Inlet Pressure	11.0 in. w.c.
Pressure Drop	0.5 in. w.c.
Specific Gravity	1.50

PIPE SIZE (in.)						
Nominal OD	$^1/_2$	$^3/_4$	1	$1^1/_4$	$1^1/_2$	2
Designation	SDR 9.33	SDR 11.0	SDR 11.00	SDR 10.00	SDR 11.00	SDR 11.00
Actual ID	0.660	0.860	1.077	1.328	1.554	1.943
Length (ft)	Capacity in Thousands of Btu per Hour					
10	340	680	1,230	2,130	3,210	5,770
20	233	468	844	1,460	2,210	3,970
30	187	375	677	1,170	1,770	3,180
40	160	321	580	1,000	1,520	2,730
50	142	285	514	890	1,340	2,420
60	129	258	466	807	1,220	2,190
70	119	237	428	742	1,120	2,010
80	110	221	398	690	1,040	1,870
90	103	207	374	648	978	1,760
100	98	196	353	612	924	1,660
125	87	173	313	542	819	1,470
150	78	157	284	491	742	1,330
175	72	145	261	452	683	1,230
200	67	135	243	420	635	1,140
250	60	119	215	373	563	1,010
300	54	108	195	338	510	916
350	50	99	179	311	469	843
400	46	92	167	289	436	784
450	43	87	157	271	409	736
500	41	82	148	256	387	695

For SI: 1 inch = 25.4 mm, 1 foot = 304.8 mm, 1 pound per square inch = 6.895 kPa, 1-inch water column = 0.2488 kPa,
1 British thermal unit per hour = 0.2931 W, 1 cubic foot per hour = 0.0283 m³/h, 1 degree = 0.01745 rad.

Note: All table entries have been rounded to three significant digits.

**TABLE 402.4(34)
POLYETHYLENE PLASTIC PIPE**

Gas	Undiluted Propane
Inlet Pressure	2.0 psi
Pressure Drop	1.0 psi
Specific Gravity	1.50

PIPE SIZE (in.)						
Nominal OD	$^1/_2$	$^3/_4$	1	$1^1/_4$	$1^1/_2$	2
Designation	SDR 9.33	SDR 11.0	SDR 11.00	SDR 10.00	SDR 11.00	SDR 11.00
Actual ID	0.660	0.860	1.077	1.328	1.554	1.943
Length (ft)	Capacity in Thousands of Btu per Hour					
10	3,130	6,260	11,300	19,600	29,500	53,100
20	2,150	4,300	7,760	13,400	20,300	36,500
30	1,730	3,450	6,230	10,800	16,300	29,300
40	1,480	2,960	5,330	9,240	14,000	25,100
50	1,310	2,620	4,730	8,190	12,400	22,200
60	1,190	2,370	4,280	7,420	11,200	20,100
70	1,090	2,180	3,940	6,830	10,300	18,500
80	1,010	2,030	3,670	6,350	9,590	17,200
90	952	1,910	3,440	5,960	9,000	16,200
100	899	1,800	3,250	5,630	8,500	15,300
125	797	1,600	2,880	4,990	7,530	13,500
150	722	1,450	2,610	4,520	6,830	12,300
175	664	1,330	2,400	4,160	6,280	11,300
200	618	1,240	2,230	3,870	5,840	10,500
250	548	1,100	1,980	3,430	5,180	9,300
300	496	994	1,790	3,110	4,690	8,430
350	457	914	1,650	2,860	4,320	7,760
400	425	851	1,530	2,660	4,020	7,220
450	399	798	1,440	2,500	3,770	6,770
500	377	754	1,360	2,360	3,560	6,390
550	358	716	1,290	2,240	3,380	6,070
600	341	683	1,230	2,140	3,220	5,790
650	327	654	1,180	2,040	3,090	5,550
700	314	628	1,130	1,960	2,970	5,330
750	302	605	1,090	1,890	2,860	5,140
800	292	585	1,050	1,830	2,760	4,960
850	283	566	1,020	1,770	2,670	4,800
900	274	549	990	1,710	2,590	4,650
950	266	533	961	1,670	2,520	4,520
1,000	259	518	935	1,620	2,450	4,400
1,100	246	492	888	1,540	2,320	4,170
1,200	234	470	847	1,470	2,220	3,980
1,300	225	450	811	1,410	2,120	3,810
1,400	216	432	779	1,350	2,040	3,660
1,500	208	416	751	1,300	1,960	3,530
1,600	201	402	725	1,260	1,900	3,410
1,700	194	389	702	1,220	1,840	3,300
1,800	188	377	680	1,180	1,780	3,200
1,900	183	366	661	1,140	1,730	3,110
2,000	178	356	643	1,110	1,680	3,020

For SI: 1 inch = 25.4 mm, 1 foot = 304.8 mm, 1 pound per square inch = 6.895 kPa, 1-inch water column = 0.2488 kPa,
          1 British thermal unit per hour = 0.2931 W, 1 cubic foot per hour = 0.0283 m³/h, 1 degree = 0.01745 rad.

Note: All table entries have been rounded to three significant digits.

TABLE 402.4(35)
POLYETHYLENE PLASTIC TUBING

Gas	Undiluted Propane
Inlet Pressure	11.0 in. w.c.
Pressure Drop	0.5 in. w.c.
Specific Gravity	1.50

	Plastic Tubing Size (CTS) (in.)	
Nominal OD	$^1/_2$	$^3/_4$
Designation	SDR 7.00	SDR 11.00
Actual ID	0.445	0.927
Length (ft)	Capacity in Cubic Feet of Gas per Hour	
10	121	828
20	83	569
30	67	457
40	57	391
50	51	347
60	46	314
70	42	289
80	39	269
90	37	252
100	35	238
125	31	211
150	28	191
175	26	176
200	24	164
225	22	154
250	21	145
275	20	138
300	19	132
350	18	121
400	16	113
450	15	106
500	15	100

For SI: 1 inch = 25.4 mm, 1 foot = 304.8 mm,
1 pound per square inch = 6.895 kPa, 1-inch water column = 0.2488 kPa,
1 British thermal unit per hour = 0.2931 W, 1 cubic foot per hour = 0.0283 m³/h,
1 degree = 0.01745 rad.

SECTION 403 (IFGS)
PIPING MATERIALS

403.1 General. Materials used for piping systems shall comply with the requirements of this chapter or shall be approved.

403.2 Used materials. Pipe, fittings, valves and other materials shall not be used again except where they are free of foreign materials and have been ascertained to be adequate for the service intended.

403.3 Other materials. Material not covered by the standards specifications listed herein shall be investigated and tested to determine that it is safe and suitable for the proposed service, and, in addition, shall be recommended for that service by the manufacturer and shall be approved by the code official.

403.4 Metallic pipe. Metallic pipe shall comply with Sections 403.4.1 through 403.4.4.

403.4.1 Cast iron. Cast-iron pipe shall not be used.

403.4.2 Steel. Steel and wrought-iron pipe shall be at least of standard weight (Schedule 40) and shall comply with one of the following standards:

1. ASME B 36.10, 10M;

2. ASTM A 53; or

3. ASTM A 106.

403.4.3 Copper and brass. Copper and brass pipe shall not be used if the gas contains more than an average of 0.3 grains of hydrogen sulfide per 100 standard cubic feet of gas (0.7 milligrams per 100 liters). Threaded copper, brass and aluminum-alloy pipe shall not be used with gases corrosive to such materials.

403.4.4 Aluminum. Aluminum-alloy pipe shall comply with ASTM B 241 (except that the use of alloy 5456 is prohibited), and shall be marked at each end of each length indicating compliance. Aluminum-alloy pipe shall be coated to protect against external corrosion where it is in contact with masonry, plaster, or insulation, or is subject to repeated wettings by such liquids as water, detergents, or sewage. Aluminum-alloy pipe shall not be used in exterior locations or underground.

403.5 Metallic tubing. Seamless copper, aluminum alloy and steel tubing shall not be used with gases corrosive to such materials.

403.5.1 Steel tubing. Steel tubing shall comply with ASTM A 254 or ASTM A 539.

403.5.2 Copper and brass tubing. Copper tubing shall comply with Standard Type K or L of ASTM B 88 or ASTM B 280.

Copper and brass tubing shall not be used if the gas contains more than an average of 0.3 grains of hydrogen sulfide per 100 standard cubic feet of gas (0.7 milligrams per 100 liters).

403.5.3 Aluminum tubing. Aluminum-alloy tubing shall comply with ASTM B 210 or ASTM B 241. Aluminum-alloy tubing shall be coated to protect against external corrosion where it is in contact with masonry, plaster or insulation, or is subject to repeated wettings by such liquids as water, detergent or sewage.

Aluminum-alloy tubing shall not be used in exterior locations or underground.

403.5.4 Corrugated stainless steel tubing. Corrugated stainless steel tubing shall be listed in accordance with ANSI LC 1/CSA 6.26.

403.6 Plastic pipe, tubing and fittings. Plastic pipe, tubing and fittings used to supply fuel gas shall be used outdoors, underground, only, and shall conform to ASTM D 2513. Pipe shall be marked "Gas" and "ASTM D 2513."

403.6.1 Anodeless risers. Plastic pipe, tubing and anodeless risers shall comply with the following:

1. Factory-assembled anodeless risers shall be recommended by the manufacturer for the gas used and shall be leak tested by the manufacturer in accordance with written procedures.

2. Service head adapters and field-assembled anodeless risers incorporating service head adapters shall be recommended by the manufacturer for the gas used, and shall be designed and certified to meet the requirements of Category I of ASTM D 2513, and U.S. Department of Transportation, Code of Federal Regulations, Title 49, Part 192.281(e). The manufacturer shall provide the user with qualified installation instructions as prescribed by the U.S. Department of Transportation, Code of Federal Regulations, Title 49, Part 192.283(b).

403.6.2 LP-gas systems. The use of plastic pipe, tubing and fittings in undiluted liquefied petroleum gas piping systems shall be in accordance with NFPA 58.

403.6.3 Regulator vent piping. Plastic pipe, tubing and fittings used to connect regulator vents to remote vent terminations shall be PVC conforming to UL 651. PVC vent piping shall not be installed indoors.

403.7 Workmanship and defects. Pipe, tubing and fittings shall be clear and free from cutting burrs and defects in structure or threading, and shall be thoroughly brushed, and chip and scale blown.

Defects in pipe, tubing and fittings shall not be repaired. Defective pipe, tubing and fittings shall be replaced (see Section 406.1.2).

403.8 Protective coating. Where in contact with material or atmosphere exerting a corrosive action, metallic piping and fittings coated with a corrosion-resistant material shall be used. External or internal coatings or linings used on piping or components shall not be considered as adding strength.

403.9 Metallic pipe threads. Metallic pipe and fitting threads shall be taper pipe threads and shall comply with ASME B1.20.1.

403.9.1 Damaged threads. Pipe with threads that are stripped, chipped, corroded or otherwise damaged shall not be used. Where a weld opens during the operation of cutting or threading, that portion of the pipe shall not be used.

403.9.2 Number of threads. Field threading of metallic pipe shall be in accordance with Table 403.9.2.

403.9.3 Thread compounds. Thread (joint) compounds (pipe dope) shall be resistant to the action of liquefied petroleum gas or to any other chemical constituents of the gases to be conducted through the piping.

403.10 Metallic piping joints and fittings. The type of piping joint used shall be suitable for the pressure-temperature conditions and shall be selected giving consideration to joint tightness and mechanical strength under the service conditions. The joint shall be able to sustain the maximum end force caused by the internal pressure and any additional forces caused by temperature expansion or contraction, vibration, fatigue or the weight of the pipe and its contents.

TABLE 403.9.2
SPECIFICATIONS FOR THREADING METALLIC PIPE

IRON PIPE SIZE (inches)	APPROXIMATE LENGTH OF THREADED PORTION (inches)	APPROXIMATE NUMBER OF THREADS TO BE CUT
$^1/_2$	$^3/_4$	10
$^3/_4$	$^3/_4$	10
1	$^7/_8$	10
$1^1/_4$	1	11
$1^1/_2$	1	11
2	1	11
$2^1/_2$	$1^1/_2$	12
3	$1^1/_2$	12
4	$1^5/_8$	13

For SI: 1 inch = 25.4 mm.

403.10.1 Pipe joints. Pipe joints shall be threaded, flanged, brazed or welded. Where nonferrous pipe is brazed, the brazing materials shall have a melting point in excess of 1,000°F (538°C). Brazing alloys shall not contain more than 0.05-percent phosphorus.

403.10.2 Tubing joints. Tubing joints shall be either made with approved gas tubing fittings or brazed with a material having a melting point in excess of 1,000°F (538°C). Brazing alloys shall not contain more than 0.05-percent phosphorus.

403.10.3 Flared joints. Flared joints shall be used only in systems constructed from nonferrous pipe and tubing where experience or tests have demonstrated that the joint is suitable for the conditions and where provisions are made in the design to prevent separation of the joints.

403.10.4 Metallic fittings. Metallic fittings shall comply with the following:

1. Threaded fittings in sizes larger than 4 inches (102 mm) shall not be used except where approved.

2. Fittings used with steel or wrought-iron pipe shall be steel, brass, bronze, malleable iron or cast iron.

3. Fittings used with copper or brass pipe shall be copper, brass or bronze.

4. Fittings used with aluminum-alloy pipe shall be of aluminum alloy.

5. Cast-iron fittings:

    5.1. Flanges shall be permitted.

    5.2. Bushings shall not be used.

    5.3. Fittings shall not be used in systems containing flammable gas-air mixtures.

    5.4. Fittings in sizes 4 inches (102 mm) and larger shall not be used indoors except where approved.

5.5. Fittings in sizes 6 inches (152 mm) and larger shall not be used except where approved.

6. Aluminum-alloy fittings. Threads shall not form the joint seal.

7. Zinc aluminum-alloy fittings. Fittings shall not be used in systems containing flammable gas-air mixtures.

8. Special fittings. Fittings such as couplings, proprietary-type joints, saddle tees, gland-type compression fittings, and flared, flareless or compression-type tubing fittings shall be: used within the fitting manufacturer's pressure-temperature recommendations; used within the service conditions anticipated with respect to vibration, fatigue, thermal expansion or contraction; installed or braced to prevent separation of the joint by gas pressure or external physical damage; and shall be approved.

403.11 Plastic pipe, joints and fittings. Plastic pipe, tubing and fittings shall be joined in accordance with the manufacturer's instructions. Such joint shall comply with the following:

1. The joint shall be designed and installed so that the longitudinal pull-out resistance of the joint will be at least equal to the tensile strength of the plastic piping material.

2. Heat-fusion joints shall be made in accordance with qualified procedures that have been established and proven by test to produce gas-tight joints at least as strong as the pipe or tubing being joined. Joints shall be made with the joining method recommended by the pipe manufacturer. Heat fusion fittings shall be marked "ASTM D 2513."

3. Where compression-type mechanical joints are used, the gasket material in the fitting shall be compatible with the plastic piping and with the gas distributed by the system. An internal tubular rigid stiffener shall be used in conjunction with the fitting. The stiffener shall be flush with the end of the pipe or tubing and shall extend at least to the outside end of the compression fitting when installed. The stiffener shall be free of rough or sharp edges and shall not be a force fit in the plastic. Split tubular stiffeners shall not be used.

4. Plastic piping joints and fittings for use in liquefied petroleum gas piping systems shall be in accordance with NFPA 58.

403.12 Flanges. All flanges shall comply with ASME B16.1, ASME B16.20 or MSS SP-6. The pressure-temperature ratings shall equal or exceed that required by the application.

403.12.1 Flange facings. Standard facings shall be permitted for use under this code. Where 150-pound (1034 kPa) pressure-rated steel flanges are bolted to Class 125 cast-iron flanges, the raised face on the steel flange shall be removed.

403.12.2 Lapped flanges. Lapped flanges shall be used only above ground or in exposed locations accessible for inspection.

403.13 Flange gaskets. Material for gaskets shall be capable of withstanding the design temperature and pressure of the piping system, and the chemical constituents of the gas being conducted, without change to its chemical and physical properties. The effects of fire exposure to the joint shall be considered in choosing material. Acceptable materials include metal or metal-jacketed asbestos (plain or corrugated), asbestos, and aluminum "O" rings and spiral wound metal gaskets. When a flanged joint is opened, the gasket shall be replaced. Full-face gaskets shall be used with all bronze and cast-iron flanges.

SECTION 404 (IFGC)
PIPING SYSTEM INSTALLATION

404.1 Prohibited locations. Piping shall not be installed in or through a circulating air duct, clothes chute, chimney or gas vent, ventilating duct, dumbwaiter or elevator shaft. Piping installed downstream of the point of delivery shall not extend through any townhouse unit other than the unit served by such piping.

404.2 Piping in solid partitions and walls. Concealed piping shall not be located in solid partitions and solid walls, unless installed in a chase or casing.

404.3 Piping in concealed locations. Portions of a piping system installed in concealed locations shall not have unions, tubing fittings, right and left couplings, bushings, compression couplings and swing joints made by combinations of fittings.

Exceptions:

1. Tubing joined by brazing.

2. Fittings listed for use in concealed locations.

404.4 Piping through foundation wall. Underground piping, where installed below grade through the outer foundation or basement wall of a building, shall be encased in a protective pipe sleeve. The annular space between the gas piping and the sleeve shall be sealed.

404.5 Protection against physical damage. In concealed locations, where piping other than black or galvanized steel is installed through holes or notches in wood studs, joists, rafters or similar members less than 1.5 inches (38 mm) from the nearest edge of the member, the pipe shall be protected by shield plates. Shield plates shall be a minimum of $^1/_{16}$-inch-thick (1.6 mm) steel, shall cover the area of the pipe where the member is notched or bored and shall extend a minimum of 4 inches (102 mm) above sole plates, below top plates and to each side of a stud, joist or rafter.

404.6 Piping in solid floors. Piping in solid floors shall be laid in channels in the floor and covered in a manner that will allow access to the piping with a minimum amount of damage to the building. Where such piping is subject to exposure to excessive moisture or corrosive substances, the piping shall be protected in an approved manner. As an alternative to installation in channels, the piping shall be installed in a conduit of Schedule 40 steel, wrought iron, PVC or ABS pipe with tightly sealed ends and joints. Both ends of such conduit shall extend not less than 2 inches (51 mm) beyond the point where the pipe emerges from the floor. The conduit shall be vented above grade to the outdoors and shall be installed so as to prevent the entry of water and insects.

404.7 Above-ground outdoor piping. All piping installed outdoors shall be elevated not less than $3^1/_2$ inches (152 mm) above ground and where installed across roof surfaces, shall be elevated not less than $3^1/_2$ inches (152 mm) above the roof surface. Piping installed above ground, outdoors, and installed across the surface of roofs shall be securely supported and located where it will be protected from physical damage. Where passing through an outside wall, the piping shall also be protected against corrosion by coating or wrapping with an inert material. Where piping is encased in a protective pipe sleeve, the annular space between the piping and the sleeve shall be sealed.

404.8 Protection against corrosion. Metallic pipe or tubing exposed to corrosive action, such as soil condition or moisture, shall be protected in an approved manner. Zinc coatings (galvanizing) shall not be deemed adequate protection for gas piping underground. Ferrous metal exposed in exterior locations shall be protected from corrosion in a manner satisfactory to the code official. Where dissimilar metals are joined underground, an insulating coupling or fitting shall be used. Piping shall not be laid in contact with cinders.

404.8.1 Prohibited use. Uncoated threaded or socket welded joints shall not be used in piping in contact with soil or where internal or external crevice corrosion is known to occur.

404.8.2 Protective coatings and wrapping. Pipe protective coatings and wrappings shall be approved for the application and shall be factory applied.

Exception: Where installed in accordance with the manufacturer's installation instructions, field application of coatings and wrappings shall be permitted for pipe nipples, fittings and locations where the factory coating or wrapping has been damaged or necessarily removed at joints.

404.9 Minimum burial depth. Underground piping systems shall be installed a minimum depth of 12 inches (305 mm) below grade, except as provided for in Section 404.9.1.

404.9.1 Individual outside appliances. Individual lines to outside lights, grills or other appliances shall be installed a minimum of 8 inches (203 mm) below finished grade, provided that such installation is approved and is installed in locations not susceptible to physical damage.

404.10 Trenches. The trench shall be graded so that the pipe has a firm, substantially continuous bearing on the bottom of the trench.

404.11 Piping underground beneath buildings. Piping installed underground beneath buildings is prohibited except where the piping is encased in a conduit of wrought iron, plastic pipe, or steel pipe designed to withstand the superimposed loads. Such conduit shall extend into an occupiable portion of the building and, at the point where the conduit terminates in the building, the space between the conduit and the gas piping shall be sealed to prevent the possible entrance of any gas leakage. Where the end sealing is capable of withstanding the full pressure of the gas pipe, the conduit shall be designed for the same pressure as the pipe. Such conduit shall extend not less than 4 inches (102 mm) outside the building, shall be vented

above grade to the outdoors, and shall be installed so as to prevent the entrance of water and insects. The conduit shall be protected from corrosion in accordance with Section 404.8.

404.12 Outlet closures. Gas outlets that do not connect to appliances shall be capped gas tight.

Exception: Listed and labeled flush-mounted-type quick-disconnect devices and listed and labeled gas convenience outlets shall be installed in accordance with the manufacturer's installation instructions.

404.13 Location of outlets. The unthreaded portion of piping outlets shall extend not less than 1 inch (25 mm) through finished ceilings and walls and where extending through floors or outdoor patios and slabs, shall not be less than 2 inches (51 mm) above them. The outlet fitting or piping shall be securely supported. Outlets shall not be placed behind doors. Outlets shall be located in the room or space where the appliance is installed.

Exception: Listed and labeled flush-mounted-type quick-disconnect devices and listed and labeled gas convenience outlets shall be installed in accordance with the manufacturer's installation instructions.

404.14 Plastic pipe. The installation of plastic pipe shall comply with Sections 404.14.1 through 404.14.3.

404.14.1 Limitations. Plastic pipe shall be installed outside underground only. Plastic pipe shall not be used within or under any building or slab or be operated at pressures greater than 100 psig (689 kPa) for natural gas or 30 psig (207 kPa) for LP-gas.

Exceptions:

1. Plastic pipe shall be permitted to terminate above ground outside of buildings where installed in premanufactured anodeless risers or service head adapter risers that are installed in accordance with the manufacturer's installation instructions.

2. Plastic pipe shall be permitted to terminate with a wall head adapter within buildings where the plastic pipe is inserted in a piping material for fuel gas use in buildings.

404.14.2 Connections. Connections made outside and underground between metallic and plastic piping shall be made only with transition fittings categorized as Category I in accordance with ASTM D 2513.

404.14.3 Tracer. A yellow insulated copper tracer wire or other approved conductor shall be installed adjacent to underground nonmetallic piping. Access shall be provided to the tracer wire or the tracer wire shall terminate above ground at each end of the nonmetallic piping. The tracer wire size shall not be less than 18 AWG and the insulation type shall be suitable for direct burial.

404.15 Prohibited devices. A device shall not be placed inside the piping or fittings that will reduce the cross-sectional area or otherwise obstruct the free flow of gas.

Exception: Approved gas filters.

404.16 Testing of piping. Before any system of piping is put in service or concealed, it shall be tested to ensure that it is gas tight. Testing, inspection and purging of piping systems shall comply with Section 406.

SECTION 405 (IFGS)
PIPING BENDS AND CHANGES IN DIRECTION

405.1 General. Changes in direction of pipe shall be permitted to be made by the use of fittings, factory bends, or field bends.

405.2 Metallic pipe. Metallic pipe bends shall comply with the following:

1. Bends shall be made only with bending tools and procedures intended for that purpose.

2. All bends shall be smooth and free from buckling, cracks or other evidence of mechanical damage.

3. The longitudinal weld of the pipe shall be near the neutral axis of the bend.

4. Pipe shall not be bent through an arc of more than 90 degrees (1.6 rad).

5. The inside radius of a bend shall be not less than six times the outside diameter of the pipe.

405.3 Plastic pipe. Plastic pipe bends shall comply with the following:

1. The pipe shall not be damaged and the internal diameter of the pipe shall not be effectively reduced.

2. Joints shall not be located in pipe bends.

3. The radius of the inner curve of such bends shall not be less than 25 times the inside diameter of the pipe.

4. Where the piping manufacturer specifies the use of special bending tools or procedures, such tools or procedures shall be used.

405.4 Elbows. Factory-made welding elbows or transverse segments cut therefrom shall have an arc length measured along the crotch at least 1 inch (25 mm) in pipe sizes 2 inches (51 mm) and larger.

SECTION 406 (IFGS)
INSPECTION, TESTING AND PURGING

406.1 General. Prior to acceptance and initial operation, all piping installations shall be inspected and pressure tested to determine that the materials, design, fabrication, and installation practices comply with the requirements of this code.

406.1.1 Inspections. Inspection shall consist of visual examination, during or after manufacture, fabrication, assembly, or pressure tests as appropriate. Supplementary types of nondestructive inspection techniques, such as magnetic-particle, radiographic, ultrasonic, etc., shall not be required unless specifically listed herein or in the engineering design.

406.1.2 Repairs and additions. In the event repairs or additions are made after the pressure test, the affected piping shall be tested.

Minor repairs and additions are not required to be pressure tested provided that the work is inspected and connec-

tions are tested with a noncorrosive leak-detecting fluid or other approved leak-detecting methods.

406.1.3 New branches. Where new branches are installed to new appliances, only the newly installed branches shall be required to be pressure tested. Connections between the new piping and the existing piping shall be tested with a noncorrosive leak-detecting fluid or other approved leak-detecting methods.

406.1.4 Section testing. A piping system shall be permitted to be tested as a complete unit or in sections. Under no circumstances shall a valve in a line be used as a bulkhead between gas in one section of the piping system and test medium in an adjacent section, unless two valves are installed in series with a valved "telltale" located between these valves. A valve shall not be subjected to the test pressure unless it can be determined that the valve, including the valve-closing mechanism, is designed to safely withstand the test pressure.

406.1.5 Regulators and valve assemblies. Regulator and valve assemblies fabricated independently of the piping system in which they are to be installed shall be permitted to be tested with inert gas or air at the time of fabrication.

406.2 Test medium. The test medium shall be air, nitrogen, carbon dioxide or an inert gas. Oxygen shall not be used.

406.3 Test preparation. Pipe joints, including welds, shall be left exposed for examination during the test.

Exception: Covered or concealed pipe end joints that have been previously tested in accordance with this code.

406.3.1 Expansion joints. Expansion joints shall be provided with temporary restraints, if required, for the additional thrust load under test.

406.3.2 Appliance and equipment isolation. Appliances and equipment that are not to be included in the test shall be either disconnected from the piping or isolated by blanks, blind flanges, or caps. Flanged joints at which blinds are inserted to blank off other equipment during the test shall not be required to be tested.

406.3.3 Appliance and equipment disconnection. Where the piping system is connected to appliances or equipment designed for operating pressures of less than the test pressure, such appliances or equipment shall be isolated from the piping system by disconnecting them and capping the outlet(s).

406.3.4 Valve isolation. Where the piping system is connected to appliances or equipment designed for operating pressures equal to or greater than the test pressure, such appliances or equipment shall be isolated from the piping system by closing the individual appliance or equipment shutoff valve(s).

406.3.5 Testing precautions. All testing of piping systems shall be done with due regard for the safety of employees and the public during the test. Bulkheads, anchorage, and bracing suitably designed to resist test pressures shall be installed if necessary. Prior to testing, the interior of the pipe shall be cleared of all foreign material.

406.4 Test pressure measurement. Test pressure shall be measured with a manometer or with a pressure-measuring device designed and calibrated to read, record, or indicate a pressure loss caused by leakage during the pressure test period. The source of pressure shall be isolated before the pressure tests are made. Mechanical gauges used to measure test pressures shall have a range such that the highest end of the scale is not greater than five times the test pressure.

406.4.1 Test pressure. The test pressure to be used shall be no less than $1^1/_2$ times the proposed maximum working pressure, but not less than 3 psig (20 kPa gauge), irrespective of design pressure. Where the test pressure exceeds 125 psig (862 kPa gauge), the test pressure shall not exceed a value that produces a hoop stress in the piping greater than 50 percent of the specified minimum yield strength of the pipe.

406.4.2 Test duration. Test duration shall be not less than $^1/_2$ hour for each 500 cubic feet (14 m³) of pipe volume or fraction thereof. When testing a system having a volume less than 10 cubic feet (0.28 m³) or a system in a single-family dwelling, the test duration shall be not less than 10 minutes. The duration of the test shall not be required to exceed 24 hours.

406.5 Detection of leaks and defects. The piping system shall withstand the test pressure specified without showing any evidence of leakage or other defects.

Any reduction of test pressures as indicated by pressure gauges shall be deemed to indicate the presence of a leak unless such reduction can be readily attributed to some other cause.

406.5.1 Detection methods. The leakage shall be located by means of an approved gas detector, a noncorrosive leak detection fluid, or other approved leak detection methods. Matches, candles, open flames, or other methods that could provide a source of ignition shall not be used.

406.5.2 Corrections. Where leakage or other defects are located, the affected portion of the piping system shall be repaired or replaced and retested.

406.6 Piping system, appliance and equipment leakage check. Leakage checking of systems and equipment shall be in accordance with Sections 406.6.1 through 406.6.4.

406.6.1 Test gases. Leak checks using fuel gas shall be permitted in piping systems that have been pressure tested in accordance with Section 406.

406.6.2 Before turning gas on. Before gas is introduced into a system of new gas piping, the entire system shall be inspected to determine that there are no open fittings or ends and that all valves at unused outlets are closed and plugged or capped.

406.6.3 Leak check. Immediately after the gas is turned on into a new system or into a system that has been initially restored after an interruption of service, the piping system shall be checked for leakage. Where leakage is indicated, the gas supply shall be shut off until the necessary repairs have been made.

406.6.4 Placing appliances and equipment in operation. Appliances and equipment shall not be placed in operation until after the piping system has been checked for leakage in accordance with Section 406.6.3 and determined to be free of leakage and purged in accordance with Section 406.7.2.

406.7 Purging. Purging of piping shall comply with Sections 406.7.1 through 406.7.4.

406.7.1 Removal from service. Where gas piping is to be opened for servicing, addition, or modification, the section to be worked on shall be turned off from the gas supply at the nearest convenient point, and the line pressure vented to the outdoors, or to ventilated areas of sufficient size to prevent accumulation of flammable mixtures.

The remaining gas in this section of pipe shall be displaced with an inert gas as required by Table 406.7.1.

TABLE 406.7.1
LENGTH OF PIPING REQUIRING PURGING WITH INERT GAS FOR SERVICING OR MODIFICATION

NOMINAL PIPE SIZE (inches)	LENGTH OF PIPING REQUIRING PURGING
$2^1/_2$	> 50 feet
3	> 30 feet
4	> 15 feet
6	> 10 feet
8 or larger	Any length

For SI: 1 inch = 25.4 mm, 1 foot = 304.8 mm.

406.7.2 Placing in operation. Where piping full of air is placed in operation, the air in the piping shall be displaced with fuel gas, except where such piping is required by Table 406.7.2 to be purged with an inert gas prior to introduction of fuel gas. The air can be safely displaced with fuel gas provided that a moderately rapid and continuous flow of fuel gas is introduced at one end of the line and air is vented out at the other end. The fuel gas flow shall be continued without interruption until the vented gas is free of air. The point of discharge shall not be left unattended during purging. After purging, the vent shall then be closed. Where required by Table 406.7.2, the air in the piping shall first be displaced with an inert gas, and the inert gas shall then be displaced with fuel gas.

TABLE 406.7.2
LENGTH OF PIPING REQUIRING PURGING WITH INERT GAS BEFORE PLACING IN OPERATION

NOMINAL PIPE SIZE (inches)	LENGTH OF PIPING REQUIRING PURGING
3	> 30 feet
4	> 15 feet
6	> 10 feet
8 or larger	Any length

For SI: 1 inch = 25.4 mm, 1 foot = 304.8 mm.

406.7.3 Discharge of purged gases. The open end of piping systems being purged shall not discharge into confined spaces or areas where there are sources of ignition unless precautions are taken to perform this operation in a safe manner by ventilation of the space, control of purging rate, and elimination of all hazardous conditions.

406.7.4 Placing appliances and equipment in operation. After the piping system has been placed in operation, all appliances and equipment shall be purged and then placed in operation, as necessary.

SECTION 407 (IFGC)
PIPING SUPPORT

407.1 General. Piping shall be provided with support in accordance with Section 407.2.

407.2 Design and installation. Piping shall be supported with pipe hooks, metal pipe straps, bands, brackets, or hangers suitable for the size of piping, of adequate strength and quality, and located at intervals so as to prevent or damp out excessive vibration. Piping shall be anchored to prevent undue strains on connected equipment and shall not be supported by other piping. Pipe hangers and supports shall conform to the requirements of MSS SP-58 and shall be spaced in accordance with Section 415. Supports, hangers, and anchors shall be installed so as not to interfere with the free expansion and contraction of the piping between anchors. All parts of the supporting equipment shall be designed and installed so they will not be disengaged by movement of the supported piping.

SECTION 408 (IFGC)
DRIPS AND SLOPED PIPING

408.1 Slopes. Piping for other than dry gas conditions shall be sloped not less than $^1/_4$ inch in 15 feet (6.3 mm in 4572 mm) to prevent traps.

408.2 Drips. Where wet gas exists, a drip shall be provided at any point in the line of pipe where condensate could collect. A drip shall also be provided at the outlet of the meter and shall be installed so as to constitute a trap wherein an accumulation of condensate will shut off the flow of gas before the condensate will run back into the meter.

408.3 Location of drips. Drips shall be provided with ready access to permit cleaning or emptying. A drip shall not be located where the condensate is subject to freezing.

408.4 Sediment trap. Where a sediment trap is not incorporated as part of the gas utilization equipment, a sediment trap shall be installed downstream of the equipment shutoff valve as close to the inlet of the equipment as practical. The sediment trap shall be either a tee fitting with a capped nipple in the bottom opening of the run of the tee or other device approved as an effective sediment trap. Illuminating appliances, ranges, clothes dryers and outdoor grills need not be so equipped.

SECTION 409 (IFGC)
SHUTOFF VALVES

409.1 General. Piping systems shall be provided with shutoff valves in accordance with this section.

409.1.1 Valve approval. Shutoff valves shall be of an approved type; shall be constructed of materials compatible with the piping; and shall comply with the standard that is applicable for the pressure and application, in accordance with Table 409.1.1.

409.1.2 Prohibited locations. Shutoff valves shall be prohibited in concealed locations and furnace plenums.

409.1.3 Access to shutoff valves. Shutoff valves shall be located in places so as to provide access for operation and shall be installed so as to be protected from damage.

409.2 Meter valve. Every meter shall be equipped with a shutoff valve located on the supply side of the meter.

409.3 Shutoff valves for multiple-house line systems. Where a single meter is used to supply gas to more than one building or tenant, a separate shutoff valve shall be provided for each building or tenant.

409.3.1 Multiple tenant buildings. In multiple tenant buildings, where a common piping system is installed to supply other than one- and two-family dwellings, shutoff valves shall be provided for each tenant. Each tenant shall have access to the shutoff valve serving that tenant's space.

409.3.2 Individual buildings. In a common system serving more than one building, shutoff valves shall be installed outdoors at each building.

409.3.3 Identification of shutoff valves. Each house line shutoff valve shall be plainly marked with an identification tag attached by the installer so that the piping systems supplied by such valves are readily identified.

409.4 MP Regulator valves. A listed shutoff valve shall be installed immediately ahead of each MP regulator.

TABLE 409.1.1
MANUAL GAS VALVE STANDARDS

VALVE STANDARDS	APPLIANCE SHUTOFF VALVE APPLICATION UP TO $^1/_2$ psig PRESSURE	OTHER VALVE APPLICATIONS			
		UP TO $^1/_2$ psig PRESSURE	UP TO 2 psig PRESSURE	UP TO 5 psig PRESSURE	UP TO 125 psig PRESSURE
ANSI Z21.15	X	—	—	—	—
CSA Requirement 3-88	X	X	X[a]	X[b]	—
ASME B16.44	X	X	X[a]	X[b]	—
ASME B16.33	X	X	X	X	X

For SI: 1 pound per square inch gauge = 6.895 kPa.
a. If labeled 2G.
b. If labeled 5G.

409.5 Equipment shutoff valve. Each appliance shall be provided with a shutoff valve separate from the appliance. The shutoff valve shall be located in the same room as the appliance, not further than 6 feet (1829 mm) from the appliance, and shall be installed upstream from the union, connector or quick disconnect device it serves. Such shutoff valves shall be provided with access.

> **Exception:** Shutoff valves for vented decorative appliances and decorative appliances for installation in vented fireplaces shall not be prohibited from being installed in an area remote from the appliance where such valves are provided with ready access. Such valves shall be permanently identified and shall serve no other equipment. Piping from the shutoff valve to within 3 feet (914 mm) of the appliance connection shall be sized in accordance with Section 402.

409.5.1 Shutoff valve in fireplace. Equipment shutoff valves located in the firebox of a fireplace shall be installed in accordance with the appliance manufacturer's instructions.

SECTION 410 (IFGC)
FLOW CONTROLS

410.1 Pressure regulators. A line pressure regulator shall be installed where the appliance is designed to operate at a lower pressure than the supply pressure. Line gas pressure regulators shall be listed as complying with ANSI Z21.80. Access shall be provided to pressure regulators. Pressure regulators shall be protected from physical damage. Regulators installed on the exterior of the building shall be approved for outdoor installation.

410.2 MP regulators. MP pressure regulators shall comply with the following:

1. The MP regulator shall be approved and shall be suitable for the inlet and outlet gas pressures for the application.

2. The MP regulator shall maintain a reduced outlet pressure under lockup (no-flow) conditions.

3. The capacity of the MP regulator, determined by published ratings of its manufacturer, shall be adequate to supply the appliances served.

4. The MP pressure regulator shall be provided with access. Where located indoors, the regulator shall be vented to the outdoors or shall be equipped with a leak-limiting device, in either case complying with Section 410.3.

5. A tee fitting with one opening capped or plugged shall be installed between the MP regulator and its upstream shutoff valve. Such tee fitting shall be positioned to allow connection of a pressure-measuring instrument and to serve as a sediment trap.

6. A tee fitting with one opening capped or plugged shall be installed not less than 10 pipe diameters downstream of the MP regulator outlet. Such tee fitting shall be positioned to allow connection of a pressure-measuring instrument.

410.3 Venting of regulators. Pressure regulators that require a vent shall be vented directly to the outdoors. The vent shall be designed to prevent the entry of insects, water and foreign objects.

> **Exception:** A vent to the outdoors is not required for regulators equipped with and labeled for utilization with an approved vent-limiting device installed in accordance with the manufacturer's instructions.

410.3.1 Vent piping. Vent piping shall be not smaller than the vent connection on the pressure regulating device. Vent piping serving relief vents and combination relief and breather vents shall be run independently to the outdoors and shall serve only a single device vent. Vent piping serving only breather vents is permitted to be connected in a manifold arrangement where sized in accordance with an approved design that minimizes back pressure in the event of diaphragm rupture.

SECTION 411 (IFGC)
APPLIANCE AND MANUFACTURED
HOME CONNECTIONS

411.1 Connecting appliances. Except as required by Section 411.1.1, appliances shall be connected to the piping system by one of the following:

1. Rigid metallic pipe and fittings.

2. Corrugated stainless steel tubing (CSST) where installed in accordance with the manufacturer's instructions.

3. Semirigid metallic tubing and metallic fittings. Lengths shall not exceed 6 feet (1829 mm) and shall be located entirely in the same room as the appliance. Semirigid metallic tubing shall not enter a motor-operated appliance through an unprotected knockout opening.

4. Listed and labeled appliance connectors in compliance with ANSI Z21.24 and installed in accordance with the manufacturer's installation instructions and located entirely in the same room as the appliance.

5. Listed and labeled quick-disconnect devices used in conjunction with listed and labeled appliance connectors.

6. Listed and labeled convenience outlets used in conjunction with listed and labeled appliance connectors.

7. Listed and labeled appliance connectors complying with ANSI Z21.69 and listed for use with food service equipment having casters, or that is otherwise subject to movement for cleaning, and other large movable equipment.

8. Listed and labeled outdoor appliance connectors in compliance with ANSI Z21.75/CSA 6.27 and installed in accordance with the manufacturer's installation instructions.

411.1.1 Commercial cooking appliances. Commercial cooking appliances that are moved for cleaning and sanitation purposes shall be connected to the piping system with an appliance connector listed as complying with ANSI Z21.69.

411.1.2 Protection against damage. Connectors and tubing shall be installed so as to be protected against physical damage.

411.1.3 Connector installation. Appliance fuel connectors shall be installed in accordance with the manufacturer's instructions and Sections 411.1.3.1 through 411.1.3.4.

411.1.3.1 Maximum length. Connectors shall have an overall length not to exceed 3 feet (914 mm), except for range and domestic clothes dryer connectors, which shall not exceed 6 feet (1829 mm) in overall length. Measurement shall be made along the centerline of the connector. Only one connector shall be used for each appliance.

> **Exception:** Rigid metallic piping used to connect an appliance to the piping system shall be permitted to have a total length greater than 3 feet (914 mm), provided that the connecting pipe is sized as part of the piping system in accordance with Section 402 and the location of the equipment shutoff valve complies with Section 409.5.

411.1.3.2 Minimum size. Connectors shall have the capacity for the total demand of the connected appliance.

411.1.3.3 Prohibited locations and penetrations. Connectors shall not be concealed within, or extended through, walls, floors, partitions, ceilings or appliance housings.

> **Exception:** Fireplace inserts that are factory equipped with grommets, sleeves or other means of protection in accordance with the listing of the appliance.

411.1.3.4 Shutoff valve. A shutoff valve not less than the nominal size of the connector shall be installed ahead of the connector in accordance with Section 409.5.

411.1.4 Movable appliances. Where appliances are equipped with casters or are otherwise subject to periodic movement or relocation for purposes such as routine cleaning and maintenance, such appliances shall be connected to the supply system piping by means of an approved flexible connector designed and labeled for the application. Such flexible connectors shall be installed and protected against physical damage in accordance with the manufacturer's installation instructions.

411.2 Manufactured home connections. Manufactured homes shall be connected to the distribution piping system by one of the following materials:

1. Metallic pipe in accordance with Section 403.4.

2. Metallic tubing in accordance with Section 403.5.

3. Listed and labeled connectors in compliance with ANSI Z21.75/CSA 6.27 and installed in accordance with the manufacturer's installation instructions.

SECTION 412 (IFGC)
LIQUEFIED PETROLEUM GAS MOTOR
VEHICLE FUEL-DISPENSING FACILITIES

[F] 412.1 General. Motor fuel-dispensing facilities for LP-gas fuel shall be in accordance with this section and the *International Fire Code*. The operation of LP-gas motor fuel-dispensing facilities shall be regulated by the *International Fire Code*.

[F] 412.2 Storage and dispensing. Storage vessels and equipment used for the storage or dispensing of LP-gas shall be approved or listed in accordance with Sections 412.3 and 412.4.

[F] 412.3 Approved equipment. Containers; pressure-relief devices, including pressure-relief valves; and pressure regulators and piping used for LP-gas shall be approved.

[F] 412.4 Listed equipment. Hoses, hose connections, vehicle fuel connections, dispensers, LP-gas pumps and electrical equipment used for LP-gas shall be listed.

[F] 412.5 Attendants. Motor vehicle fueling operations shall be conducted by qualified attendants or in accordance with Section 412.8 by persons trained in the proper handling of LP-gas.

[F] 412.6 Location. In addition to the fuel dispensing requirements of the *International Fire Code*, the point of transfer for dispensing operations shall be 25 feet (7620 mm) or more from buildings having combustible exterior wall surfaces, buildings having noncombustible exterior wall surfaces that are not part of a 1-hour fire-resistance-rated assembly or buildings having combustible overhangs, property which could be built on public streets, or sidewalks and railroads; and at least 10 feet (3048 mm) from driveways and buildings having noncombustible exterior wall surfaces that are part of a fire-resistance-rated assembly having a rating of 1 hour or more.

> **Exception:** The point of transfer for dispensing operations need not be separated from canopies providing weather protection for the dispensing equipment constructed in accordance with the *International Building Code*.

Liquefied petroleum gas containers shall be located in accordance with the *International Fire Code*. Liquefied petroleum gas storage and dispensing equipment shall be located outdoors and in accordance with the *International Fire Code*.

[F] 412.7 Installation of dispensing devices and equipment. The installation and operation of LP-gas dispensing systems shall be in accordance with this section and the *International Fire Code*. Liquefied petroleum gas dispensers and dispensing stations shall be installed in accordance with manufacturers' specifications and their listing.

[F] 412.7.1 Valves. A manual shutoff valve and an excess flow-control check valve shall be located in the liquid line between the pump and the dispenser inlet where the dispensing device is installed at a remote location and is not part of a complete storage and dispensing unit mounted on a common base.

An excess flow-control check valve or an emergency shutoff valve shall be installed in or on the dispenser at the point at which the dispenser hose is connected to the liquid piping. A differential backpressure valve shall be considered equivalent protection. A listed shutoff valve shall be located at the discharge end of the transfer hose.

[F] 412.7.2 Hoses. Hoses and piping for the dispensing of LP-gas shall be provided with hydrostatic relief valves. The hose length shall not exceed 18 feet (5486 mm). An approved method shall be provided to protect the hose against mechanical damage.

[F] 412.7.3 Vehicle impact protection. Vehicle impact protection for LP-gas storage containers, pumps and dispensers shall be provided in accordance with the *International Fire Code*.

[F] 412.8 Private fueling of motor vehicles. Self-service LP-gas dispensing systems, including key, code and card lock dispensing systems, shall not be open to the public and shall be limited to the filling of permanently mounted fuel containers on LP-gas powered vehicles. In addition to the requirements in the *International Fire Code*, self-service LP-gas dispensing systems shall be provided with an emergency shutoff switch located within 100 feet (30 480 mm) of, but not less than 20 feet (6096 mm) from, dispensers and the owner of the dispensing facility shall ensure the safe operation of the system and the training of users.

SECTION 413 (IFGC)
COMPRESSED NATURAL GAS MOTOR
VEHICLE FUEL-DISPENSING FACILITIES

[F] 413.1 General. Motor fuel-dispensing facilities for CNG fuel shall be in accordance with this section and the *International Fire Code*. The operation of CNG motor fuel-dispensing facilities shall be regulated by the *International Fire Code*.

[F] 413.2 General. Storage vessels and equipment used for the storage, compression or dispensing of CNG shall be approved or listed in accordance with Sections 413.2.1 and 413.2.3.

[F] 413.2.1 Approved equipment. Containers; compressors; pressure-relief devices, including pressure-relief valves; and pressure regulators and piping used for CNG shall be approved.

[F] 413.2.2 Listed equipment. Hoses, hose connections, dispensers, gas detection systems and electrical equipment used for CNG shall be listed. Vehicle fueling connections shall be listed and labeled.

[F] 413.2.3 General. Residential fueling appliances shall be listed. The capacity of a residential fueling appliance shall not exceed 5 standard cubic feet per minute (0.14 standard cubic meter/min) of natural gas.

[F] 413.3 Location of dispensing operations and equipment. Compression, storage and dispensing equipment shall be located above ground outside.

Exceptions:

1. Compression, storage or dispensing equipment is allowed in buildings of noncombustible construction, as set forth in the *International Building Code*, which are unenclosed for three-quarters or more of the perimeter.

2. Compression, storage and dispensing equipment is allowed to be located indoors or in vaults in accordance with the *International Fire Code*.

3. Residential fueling appliances and equipment shall be allowed to be installed indoors in accordance with the equipment manufacturer's instructions and Section 413.4.3.

[F] 413.3.1 Location on property. In addition to the fuel-dispensing requirements of the *International Fire Code*, compression, storage and dispensing equipment not located in vaults complying with the *International Fire Code* and other than residential fueling appliances shall not be installed:

1. Beneath power lines.

2. Less than 10 feet (3048 mm) from the nearest building or property line that could be built on, public street, sidewalk or source of ignition.

   Exception: Dispensing equipment need not be separated from canopies that provide weather protection for the dispensing equipment and are constructed in accordance with the *International Building Code*.

3. Less than 25 feet (7620 mm) from the nearest rail of any railroad track.

4. Less than 50 feet (15 240 mm) from the nearest rail of any railroad main track or any railroad or transit line where power for train propulsion is provided by an outside electrical source, such as third rail or overhead catenary.

5. Less than 50 feet (15 240 mm) from the vertical plane below the nearest overhead wire of a trolley bus line.

[F] 413.4 Residential fueling appliance installation. Residential fueling appliances shall be installed in accordance with Sections 413.4.1 through 413.4.3.

[F] 413.4.1 Gas connections. Residential fueling appliances shall be connected to the premises, gas piping system without causing damage to the piping system or the connection to the internal appliance apparatus.

[F] 413.4.2 Outdoor installation. Residential fueling appliances located outdoors shall be installed on a firm, noncombustible base.

[F] 413.4.3 Indoor installation. Where located indoors, residential fueling appliances shall be vented to the outdoors. A gas detector set to operate at one-fifth of the lower limit of flammability of natural gas shall be installed in the room or space containing the appliance. The detector shall be located within 6 inches (152 mm) of the highest point in the room or space. The detector shall stop the operation of the appliance and activate an audible or a visual alarm.

[F] 413.5 Private fueling of motor vehicles. Self-service CNG-dispensing systems, including key, code and card lock dispensing systems, shall be limited to the filling of permanently mounted fuel containers on CNG-powered vehicles.

In addition to the requirements in the *International Fire Code*, the owner of a self-service CNG-dispensing facility shall ensure the safe operation of the system and the training of users.

[F] 413.6 Pressure regulators. Pressure regulators shall be designed, installed or protected so their operation will not be affected by the elements (freezing rain, sleet, snow, ice, mud or debris). This protection is allowed to be integral with the regulator.

[F] 413.7 Valves. Piping to equipment shall be provided with a remote manual shutoff valve. Such valve shall be provided with ready access.

[F] 413.8 Emergency shutdown control. An emergency shutdown device shall be located within 75 feet (22 860 mm) of, but not less than 25 feet (7620 mm) from, dispensers and shall also be provided in the compressor area. Upon activation, the emergency shutdown system shall automatically shut off the power supply to the compressor and close valves between the main gas supply and the compressor and between the storage containers and dispensers.

[F] 413.9 Discharge of CNG from motor vehicle fuel storage containers. The discharge of CNG from motor vehicle fuel cylinders for the purposes of maintenance, cylinder certification, calibration of dispensers or other activities shall be in accordance with this section. The discharge of CNG from motor vehicle fuel cylinders shall be accomplished through a closed transfer system or an approved method of atmospheric venting in accordance with Section 413.9.1 or 413.9.2.

[F] 413.9.1 Closed transfer system. A documented procedure which explains the logical sequence for discharging the cylinder shall be provided to the code official for review and approval. The procedure shall include what actions the operator will take in the event of a low-pressure or high-pressure natural gas release during the discharging activity. A drawing illustrating the arrangement of piping, regulators and equipment settings shall be provided to the code official for review and approval. The drawing shall illustrate the piping and regulator arrangement and shall be shown in spatial relation to the location of the compressor, storage vessels and emergency shutdown devices.

[F] 413.9.2 Atmospheric venting. Atmospheric venting of motor vehicle fuel cylinders shall be in accordance with Sections 413.9.2.1 through 413.9.2.6.

[F] 413.9.2.1 Plans and specifications. A drawing illustrating the location of the vessel support, piping, the method of grounding and bonding, and other requirements specified herein shall be provided to the code official for review and approval.

[F] 413.9.2.2 Cylinder stability. A method of rigidly supporting the vessel during the venting of CNG shall be provided. The selected method shall provide not less than two points of support and shall prevent the horizontal and lateral movement of the vessel. The system shall be designed to prevent the movement of the vessel based on the highest gas-release velocity through valve orifices at the vessel's rated pressure and volume. The structure or appurtenance shall be constructed of noncombustible materials.

[F] 413.9.2.3 Separation. The structure or appurtenance used for stabilizing the cylinder shall be separated from the site equipment, features and exposures and shall be located in accordance with Table 413.9.2.3.

[F] 413.9.2.4 Grounding and bonding. The structure or appurtenance used for supporting the cylinder shall be grounded in accordance with the ICC *Electrical Code.*

The cylinder valve shall be bonded prior to the commencement of venting operations.

[F] 413.9.2.5 Vent tube. A vent tube that will divert the gas flow to the atmosphere shall be installed on the cylinder prior to the commencement of the venting and purging operation. The vent tube shall be constructed of pipe or tubing materials approved for use with CNG in accordance with the *International Fire Code.*

The vent tube shall be capable of dispersing the gas a minimum of 10 feet (3048 mm) above grade level. The vent tube shall not be provided with a rain cap or other feature which would limit or obstruct the gas flow.

At the connection fitting of the vent tube and the CNG cylinder, a listed bidirectional detonation flame arrester shall be provided.

[F] 413.9.2.6 Signage. Approved NO SMOKING signs shall be posted within 10 feet (3048 mm) of the cylinder support structure or appurtenance. Approved CYLINDER SHALL BE BONDED signs shall be posted on the cylinder support structure or appurtenance.

[F] TABLE 413.9.2.3
SEPARATION DISTANCE FOR
ATMOSPHERIC VENTING OF CNG

EQUIPMENT OR FEATURE	MINIMUM SEPARATION (feet)
Buildings	25
Building openings	25
Lot lines	15
Public ways	15
Vehicles	25
CNG compressor and storage vessels	25
CNG dispensers	25

For SI: 1 foot = 304.8 mm

SECTION 414 (IFGC)
SUPPLEMENTAL AND STANDBY GAS SUPPLY

414.1 Use of air or oxygen under pressure. Where air or oxygen under pressure is used in connection with the gas supply, effective means such as a backpressure regulator and relief valve shall be provided to prevent air or oxygen from passing back into the gas piping. Where oxygen is used, installation shall be in accordance with NFPA 51.

414.2 Interconnections for standby fuels. Where supplementary gas for standby use is connected downstream from a meter or a service regulator where a meter is not provided, a device to prevent backflow shall be installed. A three-way valve installed to admit the standby supply and at the same time shut off the regular supply shall be permitted to be used for this purpose.

SECTION 415 (IFGS)
PIPING SUPPORT INTERVALS

415.1 Interval of support. Piping shall be supported at intervals not exceeding the spacing specified in Table 415.1. Spacing of supports for CSST shall be in accordance with the CSST manufacturer's instructions.

TABLE 415.1
SUPPORT OF PIPING

STEEL PIPE, NOMINAL SIZE OF PIPE (inches)	SPACING OF SUPPORTS (feet)	NOMINAL SIZE OF TUBING (SMOOTH-WALL) (inch O.D.)	SPACING OF SUPPORTS (feet)
$^{1}/_{2}$	6	$^{1}/_{2}$	4
$^{3}/_{4}$ or 1	8	$^{5}/_{8}$ or $^{3}/_{4}$	6
$1^{1}/_{4}$ or larger (horizontal)	10	$^{7}/_{8}$ or 1 (Horizontal)	8
$1^{1}/_{4}$ or larger (vertical)	Every floor level	1 or Larger (vertical)	Every floor level

For SI: 1 inch = 25.4 mm, 1 foot = 304.8 mm.

SECTION 416 (IFGS)
OVERPRESSURE PROTECTION DEVICES

416.1 General. Overpressure protection devices shall be provided in accordance with this section to prevent the pressure in the piping system from exceeding the pressure that would cause unsafe operation of any connected and properly adjusted appliances.

416.2 Protection methods. The requirements of this section shall be considered to be met and a piping system deemed to have overpressure protection where a service or line pressure regulator plus one other device are installed such that the following occur:

1. Each device limits the pressure to a value that does not exceed the maximum working pressure of the downstream system.

2. The individual failure of either device does not result in the overpressurization of the downstream system.

416.3 Device maintenance. The pressure regulating, limiting and relieving devices shall be properly maintained; and inspection procedures shall be devised or suitable instrumentation installed to detect failures or malfunctions of such devices; and replacements or repairs shall be promptly made.

416.4 Where required. A pressure-relieving or pressure-limiting device shall not be required where: (1) the gas does not contain materials that could seriously interfere with the operation of the service or line pressure regulator; (2) the operating pressure of the gas source is 60 psi (414 kPa) or less; and (3) the service or line pressure regulator has all of the following design features or characteristics:

1. Pipe connections to the service or line regulator do not exceed 2 inches (51 mm) nominal diameter.

2. The regulator is self-contained with no external static or control piping.

3. The regulator has a single port valve with an orifice diameter not greater than that recommended by the manufacturer for the maximum gas pressure at the regulator inlet.

4. The valve seat is made of resilient material designed to withstand abrasion of the gas, impurities in the gas and cutting by the valve, and to resist permanent deformation where it is pressed against the valve port.

5. The regulator is capable, under normal operating conditions, of regulating the downstream pressure within the necessary limits of accuracy and of limiting the discharge pressure under no-flow conditions to not more than 150 percent of the discharge pressure maintained under flow conditions.

416.5 Devices. Pressure-relieving or pressure-limiting devices shall be one of the following:

1. Spring-loaded relief device.

2. Pilot-loaded back pressure regulator used as a relief valve and designed so that failure of the pilot system or external control piping will cause the regulator relief valve to open.

3. A monitoring regulator installed in series with the service or line pressure regulator.

4. A series regulator installed upstream from the service or line regulator and set to continuously limit the pressure on the inlet of the service or line regulator to the maximum working pressure of the downstream piping system.

5. An automatic shutoff device installed in series with the service or line pressure regulator and set to shut off when the pressure on the downstream piping system reaches the maximum working pressure or some other predetermined pressure less than the maximum working pressure. This device shall be designed so that it will remain closed until manually reset.

6. A liquid seal relief device that can be set to open accurately and consistently at the desired pressure.

The devices shall be installed either as an integral part of the service or line pressure regulator or as separate units. Where separate pressure-relieving or pressure-limiting devices are installed, they shall comply with Sections 416.5.1 through 416.5.6.

416.5.1 Construction and installation. Pressure relieving and pressure-limiting devices shall be constructed of materials so that the operation of the devices will not be impaired by corrosion of external parts by the atmosphere or of internal parts by the gas. Pressure-relieving and pressure-limiting devices shall be designed and installed so that they can be operated to determine whether the valve is free. The devices shall also be designed and installed so that they can be tested to determine the pressure at which they will operate and examined for leakage when in the closed position.

416.5.2 External control piping. External control piping shall be protected from falling objects, excavations and other causes of damage and shall be designed and installed so that damage to any control piping will not render both the regulator and the overpressure protective device inoperative.

416.5.3 Setting. Each pressure-relieving or pressure-limiting device shall be set so that the pressure does not exceed a safe level beyond the maximum allowable working pressure for the connected piping and appliances.

416.5.4 Unauthorized operation. Precautions shall be taken to prevent unauthorized operation of any shutoff valve that will make a pressure-relieving valve or pressure-limiting device inoperative. The following are acceptable methods for complying with this provision:

1. The valve shall be locked in the open position. Authorized personnel shall be instructed in the importance of leaving the shutoff valve open and of being present while the shutoff valve is closed so that it can be locked in the open position before leaving the premises.

2. Duplicate relief valves shall be installed, each having adequate capacity to protect the system, and the isolating valves and three-way valves shall be arranged so that only one safety device can be rendered inoperative at a time.

416.5.5 Vents. The discharge stacks, vents and outlet parts of all pressure-relieving and pressure-limiting devices shall be located so that gas is safely discharged to the outdoors. Discharge stacks and vents shall be designed to prevent the entry of water, insects and other foreign material that could cause blockage. The discharge stack or vent line shall be at least the same size as the outlet of the pressure-relieving device.

416.5.6 Size of fittings, pipe and openings. The fittings, pipe and openings located between the system to be protected and the pressure-relieving device shall be sized to prevent hammering of the valve and to prevent impairment of relief capacity.

CHAPTER 5

CHIMNEYS AND VENTS

SECTION 501 (IFGC)
GENERAL

501.1 Scope. This chapter shall govern the installation, maintenance, repair and approval of factory-built chimneys, chimney liners, vents and connectors and the utilization of masonry chimneys serving gas-fired appliances. The requirements for the installation, maintenance, repair and approval of factory-built chimneys, chimney liners, vents and connectors serving appliances burning fuels other than fuel gas shall be regulated by the *International Mechanical Code*. The construction, repair, maintenance and approval of masonry chimneys shall be regulated by the *International Building Code*.

501.2 General. Every appliance shall discharge the products of combustion to the outdoors, except for appliances exempted by Section 501.8.

501.3 Masonry chimneys. Masonry chimneys shall be constructed in accordance with Section 503.5.3 and the *International Building Code*.

501.4 Minimum size of chimney or vent. Chimneys and vents shall be sized in accordance with Section 504.

501.5 Abandoned inlet openings. Abandoned inlet openings in chimneys and vents shall be closed by an approved method.

501.6 Positive pressure. Where an appliance equipped with a mechanical forced draft system creates a positive pressure in the venting system, the venting system shall be designed for positive pressure applications.

501.7 Connection to fireplace. Connection of appliances to chimney flues serving fireplaces shall be in accordance with Sections 501.7.1 through 501.7.3.

501.7.1 Closure and access. A noncombustible seal shall be provided below the point of connection to prevent entry of room air into the flue. Means shall be provided for access to the flue for inspection and cleaning.

501.7.2 Connection to factory-built fireplace flue. An appliance shall not be connected to a flue serving a factory-built fireplace unless the appliance is specifically listed for such installation. The connection shall be made in accordance with the appliance manufacturer's installation instructions.

501.7.3 Connection to masonry fireplace flue. A connector shall extend from the appliance to the flue serving a masonry fireplace such that the flue gases are exhausted directly into the flue. The connector shall be accessible or removable for inspection and cleaning of both the connector and the flue. Listed direct connection devices shall be installed in accordance with their listing.

501.8 Equipment not required to be vented. The following appliances shall not be required to be vented.

1. Ranges.

2. Built-in domestic cooking units listed and marked for optional venting.

3. Hot plates and laundry stoves.

4. Type 1 clothes dryers (Type 1 clothes dryers shall be exhausted in accordance with the requirements of Section 614).

5. A single booster-type automatic instantaneous water heater, where designed and used solely for the sanitizing rinse requirements of a dishwashing machine, provided that the heater is installed in a commercial kitchen having a mechanical exhaust system. Where installed in this manner, the draft hood, if required, shall be in place and unaltered and the draft hood outlet shall be not less than 36 inches (914 mm) vertically and 6 inches (152 mm) horizontally from any surface other than the heater.

6. Refrigerators.

7. Counter appliances.

8. Room heaters listed for unvented use.

9. Direct-fired make-up air heaters.

10. Other equipment listed for unvented use and not provided with flue collars.

11. Specialized equipment of limited input such as laboratory burners and gas lights.

Where the appliances and equipment listed in Items 5 through 11 above are installed so that the aggregate input rating exceeds 20 British thermal units (Btu) per hour per cubic feet (207 watts per m³) of volume of the room or space in which such appliances and equipment are installed, one or more shall be provided with venting systems or other approved means for conveying the vent gases to the outdoor atmosphere so that the aggregate input rating of the remaining unvented appliances and equipment does not exceed the 20 Btu per hour per cubic foot (207 watts per m³) figure. Where the room or space in which the equipment is installed is directly connected to another room or space by a doorway, archway, or other opening of comparable size that cannot be closed, the volume of such adjacent room or space shall be permitted to be included in the calculations.

501.9 Chimney entrance. Connectors shall connect to a masonry chimney flue at a point not less than 12 inches (305 mm) above the lowest portion of the interior of the chimney flue.

501.10 Connections to exhauster. Appliance connections to a chimney or vent equipped with a power exhauster shall be made on the inlet side of the exhauster. Joints on the positive pressure side of the exhauster shall be sealed to prevent flue-gas leakage as specified by the manufacturer's installation instructions for the exhauster.

501.11 Masonry chimneys. Masonry chimneys utilized to vent appliances shall be located, constructed and sized as specified in the manufacturer's installation instructions for the appliances being vented and Section 503.

501.12 Residential and low-heat appliances flue lining systems. Flue lining systems for use with residential-type and low-heat appliances shall be limited to the following:

1. Clay flue lining complying with the requirements of ASTM C 315 or equivalent. Clay flue lining shall be installed in accordance with the *International Building Code*.

2. Listed chimney lining systems complying with UL 1777.

3. Other approved materials that will resist, without cracking, softening or corrosion, flue gases and condensate at temperatures up to 1,800°F (982°C).

501.13 Category I appliance flue lining systems. Flue lining systems for use with Category I appliances shall be limited to the following:

1. Flue lining systems complying with Section 501.12.

2. Chimney lining systems listed and labeled for use with gas appliances with draft hoods and other Category I gas appliances listed and labeled for use with Type B vents.

501.14 Category II, III and IV appliance venting systems. The design, sizing and installation of vents for Category II, III and IV appliances shall be in accordance with the appliance manufacturer's installation instructions.

501.15 Existing chimneys and vents. Where an appliance is permanently disconnected from an existing chimney or vent, or where an appliance is connected to an existing chimney or vent during the process of a new installation, the chimney or vent shall comply with Sections 501.15.1 through 501.15.4.

501.15.1 Size. The chimney or vent shall be resized as necessary to control flue gas condensation in the interior of the chimney or vent and to provide the appliance or appliances served with the required draft. For Category I appliances, the resizing shall be in accordance with Section 502.

501.15.2 Flue passageways. The flue gas passageway shall be free of obstructions and combustible deposits and shall be cleaned if previously used for venting a solid or liquid fuel-burning appliance or fireplace. The flue liner, chimney inner wall or vent inner wall shall be continuous and shall be free of cracks, gaps, perforations or other damage or deterioration which would allow the escape of combustion products, including gases, moisture and creosote.

501.15.3 Cleanout. Masonry chimney flues shall be provided with a cleanout opening having a minimum height of 6 inches (152 mm). The upper edge of the opening shall be located not less than 6 inches (152 mm) below the lowest chimney inlet opening. The cleanout shall be provided with a tight-fitting, noncombustible cover.

501.15.4 Clearances. Chimneys and vents shall have airspace clearance to combustibles in accordance with the *International Building Code* and the chimney or vent manufacturer's installation instructions. Noncombustible

firestopping or fireblocking shall be provided in accordance with the *International Building Code*.

> **Exception:** Masonry chimneys equipped with a chimney lining system tested and listed for installation in chimneys in contact with combustibles in accordance with UL 1777, and installed in accordance with the manufacturer's instructions, shall not be required to have clearance between combustible materials and exterior surfaces of the masonry chimney.

SECTION 502 (IFGC)
VENTS

502.1 General. All vents, except as provided in Section 503.7, shall be listed and labeled. Type B and BW vents shall be tested in accordance with UL 441. Type L vents shall be tested in accordance with UL 641. Vents for Category II and III appliances shall be tested in accordance with UL 1738. Plastic vents for Category IV appliances shall not be required to be listed and labeled where such vents are as specified by the appliance manufacturer and are installed in accordance with the appliance manufacturer's installation instructions.

502.2 Connectors required. Connectors shall be used to connect appliances to the vertical chimney or vent, except where the chimney or vent is attached directly to the appliance. Vent connector size, material, construction and installation shall be in accordance with Section 503.

502.3 Vent application. The application of vents shall be in accordance with Table 503.4.

502.4 Insulation shield. Where vents pass through insulated assemblies, an insulation shield constructed of not less than 26 gage sheet (0.016 inch) (0.4 mm) metal shall be installed to provide clearance between the vent and the insulation material. The clearance shall not be less than the clearance to combustibles specified by the vent manufacturer's installation instructions. Where vents pass through attic space, the shield shall terminate not less than 2 inches (51 mm) above the insulation materials and shall be secured in place to prevent displacement. Insulation shields provided as part of a listed vent system shall be installed in accordance with the manufacturer's installation instructions.

502.5 Installation. Vent systems shall be sized, installed and terminated in accordance with the vent and appliance manufacturer's installation instructions and Section 503.

502.6 Support of vents. All portions of vents shall be adequately supported for the design and weight of the materials employed.

502.7 Protection against physical damage. In concealed locations, where a vent is installed through holes or notches in studs, joists, rafters or similar members less than 1.5 inches (38 mm) from the nearest edge of the member, the vent shall be protected by shield plates. Shield plates shall be a minimum of $^1/_{16}$-inch-thick (1.6 mm) steel, shall cover the area of the vent where the member is notched or bored and shall extend a minimum of 4 inches (102 mm) above sole plates, below top plates and to each side of a stud, joist or rafter.

SECTION 503 (IFGS)
VENTING OF APPLIANCES

503.1 General. This section recognizes that the choice of venting materials and the methods of installation of venting systems are dependent on the operating characteristics of the appliance being vented. The operating characteristics of vented appliances can be categorized with respect to: (1) positive or negative pressure within the venting system; and (2) whether or not the appliance generates flue or vent gases that might condense in the venting system. See Section 202 for the definitions of these vented appliance categories.

503.2 Venting systems required. Except as permitted in Sections 503.2.1 through 503.2.4 and 501.8, all appliances shall be connected to venting systems.

503.2.1 Ventilating hoods. Ventilating hoods and exhaust systems shall be permitted to be used to vent appliances installed in commercial applications (see Section 503.3.4) and to vent industrial appliances, such as where the process itself requires fume disposal.

503.2.2 Well-ventilated spaces. Where located in a large and well-ventilated space, industrial appliances shall be permitted to be operated by discharging the flue gases directly into the space.

503.2.3 Direct-vent appliances. Listed direct-vent appliances shall be installed in accordance with the manufacturer's instructions and Section 503.8, Item 3.

503.2.4 Appliances with integral vents. Appliances incorporating integral venting means shall be considered properly vented where installed in accordance with the manufacturer's instructions and Section 503.8, Items 1 and 2.

503.3 Design and construction. A venting system shall be designed and constructed so as to develop a positive flow adequate to convey flue or vent gases to the outdoors.

503.3.1 Appliance draft requirements. A venting system shall satisfy the draft requirements of the appliance in accordance with the manufacturer's instructions.

503.3.2 Design and construction. Appliances required to be vented shall be connected to a venting system designed and installed in accordance with the provisions of Sections 503.4 through 503.15.

503.3.3 Mechanical draft systems. Mechanical draft systems shall comply with the following:

1. Mechanical draft systems shall be listed and shall be installed in accordance with the manufacturer's installation instructions for both the appliance and the mechanical draft system.

2. Appliances, except incinerators, requiring venting shall be permitted to be vented by means of mechanical draft systems of either forced or induced draft design.

3. Forced draft systems and all portions of induced draft systems under positive pressure during operation shall be designed and installed so as to prevent leakage of flue or vent gases into a building.

4. Vent connectors serving appliances vented by natural draft shall not be connected into any portion of mechanical draft systems operating under positive pressure.

5. Where a mechanical draft system is employed, provisions shall be made to prevent the flow of gas to the main burners when the draft system is not performing so as to satisfy the operating requirements of the appliance for safe performance.

6. The exit terminals of mechanical draft systems shall be not less than 7 feet (2134 mm) above grade where located adjacent to public walkways and shall be located as specified in Section 503.8, Items 1 and 2.

503.3.4 Ventilating hoods and exhaust systems. Ventilating hoods and exhaust systems shall be permitted to be used to vent appliances installed in commercial applications. Where automatically operated appliances are vented through a ventilating hood or exhaust system equipped with a damper or with a power means of exhaust, provisions shall be made to allow the flow of gas to the main burners only when the damper is open to a position to properly vent the appliance and when the power means of exhaust is in operation.

503.3.5 Circulating air ducts and furnace plenums. No portion of a venting system shall extend into or pass through any circulating air duct or furnace plenum.

503.3.6 Above-ceiling air-handling spaces. Where a venting system passes through an above-ceiling air-handling space or other nonducted portion of an air-handling system, the venting system shall conform to one of the following requirements:

1. The venting system shall be a listed special gas vent; other venting system serving a Category III or Category IV appliance; or other positive pressure vent, with joints sealed in accordance with the appliance or vent manufacturer's instructions.

2. The venting system shall be installed such that fittings and joints between sections are not installed in the above-ceiling space.

3. The venting system shall be installed in a conduit or enclosure with sealed joints separating the interior of the conduit or enclosure from the ceiling space.

503.4 Type of venting system to be used. The type of venting system to be used shall be in accordance with Table 503.4.

503.4.1 Plastic piping. Plastic piping used for venting appliances listed for use with such venting materials shall be approved.

503.4.2 Special gas vent. Special gas vent shall be listed and installed in accordance with the special gas vent manufacturer's installation instructions.

TABLE 503.4
TYPE OF VENTING SYSTEM TO BE USED

APPLIANCES	TYPE OF VENTING SYSTEM
Listed Category I appliances Listed appliances equipped with draft hood Appliances listed for use with Type B gas vent	Type B gas vent (Section 503.6) Chimney (Section 503.5) Single-wall metal pipe (Section 503.7) Listed chimney lining system for gas venting (Section 503.5.3) Special gas vent listed for these appliances (Section 503.4.2)
Listed vented wall furnaces	Type B-W gas vent (Sections 503.6, 608)
Category II appliances	As specified or furnished by manufacturers of listed appliances (Sections 503.4.1, 503.4.2)
Category III appliances	As specified or furnished by manufacturers of listed appliances (Sections 503.4.1, 503.4.2)
Category IV appliances	As specified or furnished by manufacturers of listed appliances (Sections 503.4.1, 503.4.2)
Incinerators, indoors	Chimney (Section 503.5)
Incinerators, outdoors	Single-wall metal pipe (Sections 503.7, 503.7.6)
Appliances that can be converted for use with solid fuel	Chimney (Section 503.5)
Unlisted combination gas and oil-burning appliances	Chimney (Section 503.5)
Listed combination gas and oil-burning appliances	Type L vent (Section 503.6) or chimney (Section 503.5)
Combination gas and solid fuel-burning appliances	Chimney (Section 503.5)
Appliances listed for use with chimneys only	Chimney (Section 503.5)
Unlisted appliances	Chimney (Section 503.5)
Decorative appliances in vented fireplaces	Chimney
Gas-fired toilets	Single-wall metal pipe (Section 626)
Direct-vent appliances	See Section 503.2.3
Appliances with integral vent	See Section 503.2.4

503.5 Masonry, metal, and factory-built chimneys.
Masonry, metal and factory-built chimneys shall comply with Sections 503.5.1 through 503.5.10.

503.5.1 Factory-built chimneys. Factory-built chimneys shall be installed in accordance with the manufacturer's installation instructions. Factory-built chimneys used to vent appliances that operate at a positive vent pressure shall be listed for such application.

503.5.2 Metal chimneys. Metal chimneys shall be built and installed in accordance with NFPA 211.

503.5.3 Masonry chimneys. Masonry chimneys shall be built and installed in accordance with NFPA 211 and shall be lined with approved clay flue lining, a listed chimney lining system or other approved material that will resist corrosion, erosion, softening or cracking from vent gases at temperatures up to 1,800°F (982°C).

Exception: Masonry chimney flues serving listed gas appliances with draft hoods, Category I appliances and other gas appliances listed for use with Type B vents shall be permitted to be lined with a chimney lining system specifically listed for use only with such appliances. The liner shall be installed in accordance with the liner

manufacturer's installation instructions. A permanent identifying label shall be attached at the point where the connection is to be made to the liner. The label shall read: "This chimney liner is for appliances that burn gas only. Do not connect to solid or liquid fuel-burning appliances or incinerators."

For installation of gas vents in existing masonry chimneys, see Section 503.6.3.

503.5.4 Chimney termination. Chimneys for residential-type or low-heat appliances shall extend at least 3 feet (914 mm) above the highest point where they pass through a roof of a building and at least 2 feet (610 mm) higher than any portion of a building within a horizontal distance of 10 feet (3048 mm) (see Figure 503.5.4). Chimneys for medium-heat appliances shall extend at least 10 feet (3048 mm) higher than any portion of any building within 25 feet (7620 mm). Chimneys shall extend at least 5 feet (1524 mm) above the highest connected appliance draft hood outlet or flue collar. Decorative shrouds shall not be installed at the termination of factory-built chimneys except where such shrouds are listed and labeled for use with the specific factory-built chimney system and are installed in accordance with the manufacturer's installation instructions.

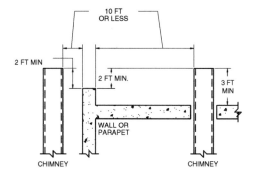

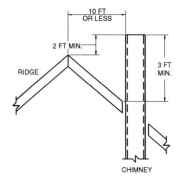

A. TERMINATION 10 FT OR LESS FROM RIDGE, WALL, OR PARAPET

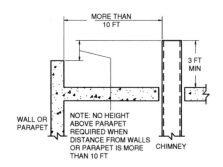

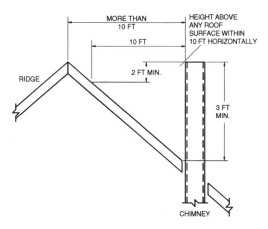

B. TERMINATION MORE THAN 10 FT FROM RIDGE, WALL, OR PARAPET

For SI: 1 inch = 25.4 mm, 1 foot = 304.8 mm.

FIGURE 503.5.4
TYPICAL TERMINATION LOCATIONS FOR
CHIMNEYS AND SINGLE-WALL METAL PIPES SERVING
RESIDENTIAL-TYPE AND LOW-HEAT EQUIPMENT

503.5.5 Size of chimneys. The effective area of a chimney venting system serving listed appliances with draft hoods, Category I appliances, and other appliances listed for use with Type B vents shall be determined in accordance with one of the following methods:

1. The provisions of Section 504.

2. For sizing an individual chimney venting system for a single appliance with a draft hood, the effective areas of the vent connector and chimney flue shall be not less than the area of the appliance flue collar or draft hood outlet, nor greater than seven times the draft hood outlet area.

3. For sizing a chimney venting system connected to two appliances with draft hoods, the effective area of the chimney flue shall be not less than the area of the larger draft hood outlet plus 50 percent of the area of the smaller draft hood outlet, nor greater than seven times the smallest draft hood outlet area.

4. Chimney venting systems using mechanical draft shall be sized in accordance with approved engineering methods.

5. Other approved engineering methods.

503.5.5.1 Incinerator venting. Where an incinerator is vented by a chimney serving other appliances, the gas input to the incinerator shall not be included in calculating chimney size, provided that the chimney flue diameter is not less than 1 inch (25 mm) larger in equivalent diameter than the diameter of the incinerator flue outlet.

503.5.6 Inspection of chimneys. Before replacing an existing appliance or connecting a vent connector to a chimney, the chimney passageway shall be examined to ascertain that it is clear and free of obstructions and it shall be cleaned if previously used for venting solid or liquid fuel-burning appliances or fireplaces.

503.5.6.1 Chimney lining. Chimneys shall be lined in accordance with NFPA 211.

Exception: Existing chimneys shall be permitted to have their use continued when an appliance is replaced by an appliance of similar type, input rating, and efficiency.

503.5.6.2 Cleanouts. Cleanouts shall be examined to determine if they will remain tightly closed when not in use.

503.5.6.3 Unsafe chimneys. Where inspection reveals that an existing chimney is not safe for the intended application, it shall be repaired, rebuilt, lined, relined or replaced with a vent or chimney to conform to NFPA 211 and it shall be suitable for the appliances to be vented.

503.5.7 Chimneys serving equipment burning other fuels. Chimneys serving equipment burning other fuels shall comply with Sections 503.5.7.1 through 503.5.7.4.

503.5.7.1 Solid fuel-burning appliances. An appliance shall not be connected to a chimney flue serving a separate appliance designed to burn solid fuel.

503.5.7.2 Liquid fuel-burning appliances. Where one chimney flue serves gas appliances and liquid fuel-burning appliances, the appliances shall be connected through separate openings or shall be connected through a single opening where joined by a suitable fitting located as close as practical to the chimney. Where two or more openings are provided into one chimney flue, they shall be at different levels. Where the appliances are automatically controlled, they shall be equipped with safety shutoff devices.

503.5.7.3 Combination gas and solid fuel-burning appliances. A combination gas- and solid fuel-burning appliance shall be permitted to be connected to a single chimney flue where equipped with a manual reset device to shut off gas to the main burner in the event of sustained backdraft or flue gas spillage. The chimney flue shall be sized to properly vent the appliance.

503.5.7.4 Combination gas- and oil fuel-burning appliances. A listed combination gas- and oil fuel-burning appliance shall be permitted to be connected to a single chimney flue. The chimney flue shall be sized to properly vent the appliance.

503.5.8 Support of chimneys. All portions of chimneys shall be supported for the design and weight of the materials employed. Factory-built chimneys shall be supported and spaced in accordance with the manufacturer's installation instructions.

503.5.9 Cleanouts. Where a chimney that formerly carried flue products from liquid or solid fuel-burning appliances is used with an appliance using fuel gas, an accessible cleanout shall be provided. The cleanout shall have a tight-fitting cover and shall be installed so its upper edge is at least 6 inches (152 mm) below the lower edge of the lowest chimney inlet opening.

503.5.10 Space surrounding lining or vent. The remaining space surrounding a chimney liner, gas vent, special gas vent or plastic piping installed within a masonry chimney flue shall not be used to vent another appliance. The insertion of another liner or vent within the chimney as provided in this code and the liner or vent manufacturer's instructions shall not be prohibited.

The remaining space surrounding a chimney liner, gas vent, special gas vent or plastic piping installed within a masonry, metal or factory-built chimney shall not be used to supply combustion air. Such space shall not be prohibited from supplying combustion air to direct-vent appliances designed for installation in a solid fuel-burning fireplace and installed in accordance with the manufacturer's installation instructions.

503.6 Gas vents. Gas vents shall comply with Sections 503.6.1 through 503.6.12 (see Section 202, Definitions).

503.6.1 Installation, general. Gas vents shall be installed in accordance with the manufacturer's installation instructions.

503.6.2 Type B-W vent capacity. A Type B-W gas vent shall have a listed capacity not less than that of the listed vented wall furnace to which it is connected.

503.6.3 Gas vents installed within masonry chimneys. Gas vents installed within masonry chimneys shall be installed in accordance with the manufacturer's installation instructions. Gas vents installed within masonry chimneys shall be identified with a permanent label installed at the point where the vent enters the chimney. The label shall contain the following language: "This gas vent is for appliances that burn gas. Do not connect to solid or liquid fuel-burning appliances or incinerators."

503.6.4 Gas vent terminations. A gas vent shall terminate in accordance with one of the following:

1. Gas vents that are 12 inches (305 mm) or less in size and located not less than 8 feet (2438 mm) from a vertical wall or similar obstruction shall terminate above the roof in accordance with Figure 503.6.4.

2. Gas vents that are over 12 inches (305 mm) in size or are located less than 8 feet (2438 mm) from a vertical wall or similar obstruction shall terminate not less than 2 feet (610 mm) above the highest point where they pass through the roof and not less than 2 feet (610 mm) above any portion of a building within 10 feet (3048 mm) horizontally.

3. As provided for industrial appliances in Section 503.2.2.

4. As provided for direct-vent systems in Section 503.2.3.

5. As provided for appliances with integral vents in Section 503.2.4.

6. As provided for mechanical draft systems in Section 503.3.3.

7. As provided for ventilating hoods and exhaust systems in Section 503.3.4.

503.6.4.1 Decorative shrouds. Decorative shrouds shall not be installed at the termination of gas vents except where such shrouds are listed for use with the specific gas venting system and are installed in accordance with manufacturer's installation instructions.

503.6.5 Minimum height. A Type B or L gas vent shall terminate at least 5 feet (1524 mm) in vertical height above the highest connected appliance draft hood or flue collar. A Type B-W gas vent shall terminate at least 12 feet (3658 mm) in vertical height above the bottom of the wall furnace.

503.6.6 Roof terminations. Gas vents shall extend through the roof flashing, roof jack or roof thimble and terminate with a listed cap or listed roof assembly.

503.6.7 Forced air inlets. Gas vents shall terminate not less than 3 feet (914 mm) above any forced air inlet located within 10 feet (3048 mm).

503.6.8 Exterior wall penetrations. A gas vent extending through an exterior wall shall not terminate adjacent to the wall or below eaves or parapets, except as provided in Sections 503.2.3 and 503.3.3.

503.6.9 Size of gas vents. Venting systems shall be sized and constructed in accordance with Section 504 or other approved engineering methods and the gas vent and appliance manufacturer's installation instructions.

503.6.9.1 Category I appliances. The sizing of natural draft venting systems serving one or more listed appliances equipped with a draft hood or appliances listed for use with Type B gas vent, installed in a single story of a building, shall be in accordance with one of the following methods:

1. The provisions of Section 504.

2. For sizing an individual gas vent for a single, draft-hood-equipped appliance, the effective area of the vent connector and the gas vent shall be not less than the area of the appliance draft hood outlet, nor greater than seven times the draft hood outlet area.

3. For sizing a gas vent connected to two appliances with draft hoods, the effective area of the vent shall

be not less than the area of the larger draft hood outlet plus 50 percent of the area of the smaller draft hood outlet, nor greater than seven times the smaller draft hood outlet area.

4. Approved engineering practices.

503.6.9.2 Vent offsets. Type B and L vents sized in accordance with Item 2 or 3 of Section 503.6.9.1 shall extend in a generally vertical direction with offsets not exceeding 45 degrees (0.79 rad), except that a vent system having not more than one 60-degree (1.04 rad) offset shall be permitted. Any angle greater than 45 degrees (0.79 rad) from the vertical is considered horizontal. The total horizontal distance of a vent plus the horizontal vent connector serving draft hood-equipped appliances shall be not greater than 75 percent of the vertical height of the vent.

503.6.9.3 Category II, III and IV appliances. The sizing of gas vents for Category II, III and IV appliances shall be in accordance with the appliance manufacturer's instructions.

503.6.9.4 Mechanical draft. Chimney venting systems using mechanical draft shall be sized in accordance with approved engineering methods.

503.6.10 Gas vents serving appliances on more than one floor. A common gas vent shall be permitted in multistory installations to vent Category I appliances located on more than one floor level, provided that the venting system is designed and installed in accordance with approved engineering methods. For the purpose of this section, crawl spaces, basements and attics shall be considered as floor levels.

503.6.10.1 Appliance separation. All appliances connected to the common vent shall be located in rooms separated from occupiable space. Each of these rooms shall have provisions for an adequate supply of combustion, ventilation and dilution air that is not supplied from an occupiable space (see Figure 503.6.10.1).

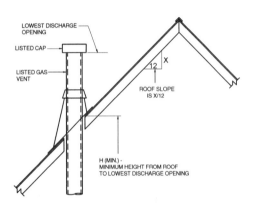

ROOF SLOPE	H (min) ft
Flat to 6/12	1.0
Over 6/12 to 7/12	1.25
Over 7/12 to 8/12	1.5
Over 8/12 to 9/12	2.0
Over 9/12 to 10/12	2.5
Over 10/12 to 11/12	3.25
Over 11/12 to 12/12	4.0
Over 12/12 to 14/12	5.0
Over 14/12 to 16/12	6.0
Over 16/12 to 18/12	7.0
Over 18/12 to 20/12	7.5
Over 20/12 to 21/12	8.0

For SI: 1 inch = 25.4 mm, 1 foot = 304.8 mm.

**FIGURE 503.6.4
TERMINATION LOCATIONS FOR GAS VENTS WITH
LISTED CAPS 12 INCHES OR LESS IN SIZE AT LEAST 8 FEET
FROM A VERTICAL WALL**

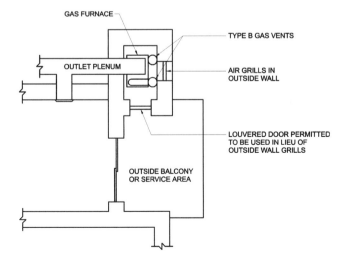

**FIGURE 503.6.10.1
PLAN VIEW OF PRACTICAL SEPARATION METHOD
FOR MULTISTORY GAS VENTING**

503.6.10.2 Sizing. The size of the connectors and common segments of multistory venting systems for appliances listed for use with Type B double-wall gas vents shall be in accordance with Table 504.3(1), provided that:

1. The available total height (*H*) for each segment of a multistory venting system is the vertical distance between the level of the highest draft hood outlet or flue collar on that floor and the centerline of the next highest interconnection tee (see Figure B-13).

2. The size of the connector for a segment is determined from the appliance input rating and available connector rise, and shall not be smaller than the draft hood outlet or flue collar size.

3. The size of the common vertical segment, and of the interconnection tee at the base of that segment, shall be based on the total appliance input rating entering that segment and its available total height.

503.6.11 Support of gas vents. Gas vents shall be supported and spaced in accordance with the manufacturer's installation instructions.

503.6.12 Marking. In those localities where solid and liquid fuels are used extensively, gas vents shall be permanently identified by a label attached to the wall or ceiling at a point where the vent connector enters the gas vent. The determination of where such localities exist shall be made by the code official. The label shall read:

"This gas vent is for appliances that burn gas. Do not connect to solid or liquid fuel-burning appliances or incinerators."

503.7 Single-wall metal pipe. Single-wall metal pipe vents shall comply with Sections 503.7.1 through 503.7.12.

503.7.1 Construction. Single-wall metal pipe shall be constructed of galvanized sheet steel not less than 0.0304 inch (0.7 mm) thick, or other approved, noncombustible, corrosion-resistant material.

503.7.2 Cold climate. Uninsulated single-wall metal pipe shall not be used outdoors for venting appliances in regions where the 99-percent winter design temperature is below 32°F (0°C).

503.7.3 Termination. Single-wall metal pipe shall terminate at least 5 feet (1524 mm) in vertical height above the highest connected appliance draft hood outlet or flue collar. Single-wall metal pipe shall extend at least 2 feet (610 mm) above the highest point where it passes through a roof of a building and at least 2 feet (610 mm) higher than any portion of a building within a horizontal distance of 10 feet (3048 mm) (see Figure 503.5.4). An approved cap or roof assembly shall be attached to the terminus of a single-wall metal pipe (see also Section 503.7.8, Item 3).

503.7.4 Limitations of use. Single-wall metal pipe shall be used only for runs directly from the space in which the appliance is located through the roof or exterior wall to the outdoor atmosphere.

503.7.5 Roof penetrations. A pipe passing through a roof shall extend without interruption through the roof flashing, roof jack, or roof thimble. Where a single-wall metal pipe passes through a roof constructed of combustible material, a noncombustible, nonventilating thimble shall be used at the point of passage. The thimble shall extend at least 18 inches (457 mm) above and 6 inches (152 mm) below the roof with the annular space open at the bottom and closed only at the top. The thimble shall be sized in accordance with Section 503.10.15.

503.7.6 Installation. Single-wall metal pipe shall not originate in any unoccupied attic or concealed space and shall not pass through any attic, inside wall, concealed space, or floor. The installation of a single-wall metal pipe through an exterior combustible wall shall comply with Section 503.10.15. Single-wall metal pipe used for venting an incinerator shall be exposed and readily examinable for its full length and shall have suitable clearances maintained.

503.7.7 Clearances. Minimum clearances from single-wall metal pipe to combustible material shall be in accordance with Table 503.7.7. The clearance from single-wall metal pipe to combustible material shall be permitted to be reduced where the combustible material is protected as specified for vent connectors in Table 308.2.

503.7.8 Size of single-wall metal pipe. A venting system constructed of single-wall metal pipe shall be sized in accordance with one of the following methods and the appliance manufacturer's instructions:

1. For a draft-hood-equipped appliance, in accordance with Section 504.

2. For a venting system for a single appliance with a draft hood, the areas of the connector and the pipe each shall be not less than the area of the appliance flue collar or draft hood outlet, whichever is smaller. The vent area shall not be greater than seven times the draft hood outlet area.

3. Other approved engineering methods.

503.7.9 Pipe geometry. Any shaped single-wall metal pipe shall be permitted to be used, provided that its equivalent effective area is equal to the effective area of the round pipe for which it is substituted, and provided that the minimum internal dimension of the pipe is not less than 2 inches (51 mm).

503.7.10 Termination capacity. The vent cap or a roof assembly shall have a venting capacity not less than that of the pipe to which it is attached.

503.7.11 Support of single-wall metal pipe. All portions of single-wall metal pipe shall be supported for the design and weight of the material employed.

503.7.12 Marking. Single-wall metal pipe shall comply with the marking provisions of Section 503.6.12.

TABLE 503.7.7ª
CLEARANCES FOR CONNECTORS

APPLIANCE	MINIMUM DISTANCE FROM COMBUSTIBLE MATERIAL			
	Listed Type B gas vent material	Listed Type L vent material	Single-wall metal pipe	Factory-built chimney sections
Listed appliances with draft hoods and appliances listed for use with Type B gas vents	As listed	As listed	6 inches	As listed
Residential boilers and furnaces with listed gas conversion burner and with draft hood	6 inches	6 inches	9 inches	As listed
Residential appliances listed for use with Type L vents	Not permitted	As listed	9 inches	As listed
Listed gas-fired toilets	Not permitted	As listed	As listed	As listed
Unlisted residential appliances with draft hood	Not permitted	6 inches	9 inches	As listed
Residential and low-heat appliances other than above	Not permitted	9 inches	18 inches	As listed
Medium-heat appliances	Not permitted	Not permitted	36 inches	As listed

For SI: 1 inch = 25.4 mm.

a. These clearances shall apply unless the manufacturer's installation instructions for a listed appliance or connector specify different clearances, in which case the listed clearances shall apply.

503.8 Venting system termination location. The location of venting system terminations shall comply with the following (see Appendix C):

1. A mechanical draft venting system shall terminate at least 3 feet (914 mm) above any forced-air inlet located within 10 feet (3048 mm).

   Exceptions:

   1. This provision shall not apply to the combustion air intake of a direct-vent appliance.

   2. This provision shall not apply to the separation of the integral outdoor air inlet and flue gas discharge of listed outdoor appliances.

2. A mechanical draft venting system, excluding direct-vent appliances, shall terminate at least 4 feet (1219 mm) below, 4 feet (1219 mm) horizontally from, or 1 foot (305 mm) above any door, operable window, or gravity air inlet into any building. The bottom of the vent terminal shall be located at least 12 inches (305 mm) above grade.

3. The vent terminal of a direct-vent appliance with an input of 10,000 Btu per hour (3 kW) or less shall be located at least 6 inches (152 mm) from any air opening into a building, and such an appliance with an input over 10,000 Btu per hour (3 kW) but not over 50,000 Btu per hour (14.7 kW) shall be installed with a 9-inch (230 mm) vent termination clearance, and an appliance with an input over 50,000 Btu/h (14.7 kw) shall have at least a 12-inch (305 mm) vent termination clearance. The bottom of the vent terminal and the air intake shall be located at least 12 inches (305 mm) above grade.

4. Through-the-wall vents for Category II and IV appliances and noncategorized condensing appliances shall not terminate over public walkways or over an area where condensate or vapor could create a nuisance or hazard or could be detrimental to the operation of regulators, relief valves, or other equipment. Where local experience indicates that condensate is a problem with Category I and III appliances, this provision shall also apply.

503.9 Condensation drainage. Provisions shall be made to collect and dispose of condensate from venting systems serving Category II and IV appliances and noncategorized condensing appliances in accordance with Section 503.8, Item 4. Where local experience indicates that condensation is a problem, provision shall be made to drain off and dispose of condensate from venting systems serving Category I and III appliances in accordance with Section 503.8, Item 4.

503.10 Vent connectors for Category I equipment. Vent connectors for Category I equipment shall comply with Sections 503.10.1 through 503.10.16.

503.10.1 Where required. A vent connector shall be used to connect an appliance to a gas vent, chimney or single-wall metal pipe, except where the gas vent, chimney or single-wall metal pipe is directly connected to the appliance.

503.10.2 Materials. Vent connectors shall be constructed in accordance with Sections 503.10.2.1 through 503.10.2.5.

503.10.2.1 General. A vent connector shall be made of noncombustible corrosion-resistant material capable of withstanding the vent gas temperature produced by the appliance and of sufficient thickness to withstand physical damage.

503.10.2.2 Vent connectors located in unconditioned areas. Where the vent connector used for an appliance having a draft hood or a Category I appliance is located in or passes through attics, crawl spaces or other unconditioned spaces, that portion of the vent connector shall be listed Type B, Type L or listed vent material having equivalent insulation properties.

Exception: Single-wall metal pipe located within the exterior walls of the building in areas having a local 99-percent winter design temperature of 5°F (-15°C) or higher shall be permitted to be used in unconditioned spaces other than attics and crawl spaces.

503.10.2.3 Residential-type appliance connectors. Where vent connectors for residential-type appliances are not installed in attics or other unconditioned spaces, connectors for listed appliances having draft hoods, appliances having draft hoods and equipped with listed conversion burners and Category I appliances shall be one of the following:

1. Type B or L vent material;

2. Galvanized sheet steel not less than 0.018 inch (0.46 mm) thick;

3. Aluminum (1100 or 3003 alloy or equivalent) sheet not less than 0.027 inch (0.69 mm) thick;

4. Stainless steel sheet not less than 0.012 inch (0.31 mm) thick;

5. Smooth interior wall metal pipe having resistance to heat and corrosion equal to or greater than that of Item 2, 3 or 4 above; or

6. A listed vent connector.

Vent connectors shall not be covered with insulation.

Exception: Listed insulated vent connectors shall be installed according to the terms of their listing.

503.10.2.4 Low-heat equipment. A vent connector for a nonresidential, low-heat appliance shall be a factory-built chimney section or steel pipe having resistance to heat and corrosion equivalent to that for the appropriate galvanized pipe as specified in Table 503.10.2.4. Factory-built chimney sections shall be joined together in accordance with the chimney manufacturer's instructions.

TABLE 503.10.2.4
MINIMUM THICKNESS FOR GALVANIZED STEEL VENT CONNECTORS FOR LOW-HEAT APPLIANCES

DIAMETER OF CONNECTOR (inches)	MINIMUM THICKNESS (inch)
Less than 6	0.019
6 to less than 10	0.023
10 to 12 inclusive	0.029
14 to 16 inclusive	0.034
Over 16	0.056

For SI: 1 inch = 25.4 mm.

503.10.2.5 Medium-heat appliances. Vent connectors for medium-heat appliances and commercial and industrial incinerators shall be constructed of factory-built medium-heat chimney sections or steel of a thickness not less than that specified in Table 503.10.2.5 and shall comply with the following:

1. A steel vent connector for an appliance with a vent gas temperature in excess of 1,000°F (538°C) measured at the entrance to the connector shall be lined with medium-duty fire brick (ASTM C 64, Type F), or the equivalent.

2. The lining shall be at least $2^1/_2$ inches (64 mm) thick for a vent connector having a diameter or

greatest cross-sectional dimension of 18 inches (457 mm) or less.

3. The lining shall be at least $4^1/_2$ inches (114 mm) thick laid on the $4^1/_2$-inch (114 mm) bed for a vent connector having a diameter or greatest cross-sectional dimension greater than 18 inches (457 mm).

4. Factory-built chimney sections, if employed, shall be joined together in accordance with the chimney manufacturer's instructions.

TABLE 503.10.2.5
MINIMUM THICKNESS FOR STEEL VENT CONNECTORS FOR MEDIUM-HEAT APPLIANCES AND COMMERCIAL AND INDUSTRIAL INCINERATORS VENT CONNECTOR SIZE

DIAMETER (inches)	AREA (square inches)	MINIMUM THICKNESS (inch)
Up to 14	Up to 154	0.053
Over 14 to 16	154 to 201	0.067
Over 16 to 18	201 to 254	0.093
Over 18	Larger than 254	0.123

For SI: 1 inch = 25.4 mm, 1 square inch = 645.16 mm².

503.10.3 Size of vent connector. Vent connectors shall be sized in accordance with Sections 503.10.3.1 through 503.10.3.5.

503.10.3.1 Single draft hood and fan-assisted. A vent connector for an appliance with a single draft hood or for a Category I fan-assisted combustion system appliance shall be sized and installed in accordance with Section 504 or other approved engineering methods.

503.10.3.2 Multiple draft hood. For a single appliance having more than one draft hood outlet or flue collar, the manifold shall be constructed according to the instructions of the appliance manufacturer. Where there are no instructions, the manifold shall be designed and constructed in accordance with approved engineering practices. As an alternate method, the effective area of the manifold shall equal the combined area of the flue collars or draft hood outlets and the vent connectors shall have a minimum 1-foot (305 mm) rise.

503.10.3.3 Multiple appliances. Where two or more appliances are connected to a common vent or chimney, each vent connector shall be sized in accordance with Section 504 or other approved engineering methods.

As an alternative method applicable only when all of the appliances are draft hood equipped, each vent connector shall have an effective area not less than the area of the draft hood outlet of the appliance to which it is connected.

503.10.3.4 Common connector/manifold. Where two or more appliances are vented through a common vent connector or vent manifold, the common vent connector or vent manifold shall be located at the highest level consistent with available headroom and the required clearance to combustible materials and shall be sized in accordance with Section 504 or other approved engineering methods.

As an alternate method applicable only where there are two draft hood-equipped appliances, the effective area of the common vent connector or vent manifold and all junction fittings shall be not less than the area of the larger vent connector plus 50 percent of the area of the smaller flue collar outlet.

503.10.3.5 Size increase. Where the size of a vent connector is increased to overcome installation limitations and obtain connector capacity equal to the appliance input, the size increase shall be made at the appliance draft hood outlet.

503.10.4 Two or more appliances connected to a single vent. Where two or more vent connectors enter a common gas vent, chimney flue, or single-wall metal pipe, the smaller connector shall enter at the highest level consistent with the available headroom or clearance to combustible material. Vent connectors serving Category I appliances shall not be connected to any portion of a mechanical draft system operating under positive static pressure, such as those serving Category III or IV appliances.

503.10.5 Clearance. Minimum clearances from vent connectors to combustible material shall be in accordance with Table 503.7.7.

> **Exception:** The clearance between a vent connector and combustible material shall be permitted to be reduced where the combustible material is protected as specified for vent connectors in Table 308.2.

503.10.6 Flow resistance. A vent connector shall be installed so as to avoid turns or other construction features that create excessive resistance to flow of vent gases.

503.10.7 Joints. Joints between sections of connector piping and connections to flue collars and draft hood outlets shall be fastened by one of the following methods:

1. Sheet metal screws.

2. Vent connectors of listed vent material assembled and connected to flue collars or draft hood outlets in accordance with the manufacturers' instructions.

3. Other approved means.

503.10.8 Slope. A vent connector shall be installed without dips or sags and shall slope upward toward the vent or chimney at least $^1/_4$ inch per foot (21 mm/m).

> **Exception:** Vent connectors attached to a mechanical draft system installed in accordance with the manufacturers' instructions.

503.10.9 Length of vent connector. A vent connector shall be as short as practical and the appliance located as close as practical to the chimney or vent. The maximum horizontal length of a single-wall connector shall be 75 percent of the height of the chimney or vent except for engineered systems. The maximum horizontal length of a Type B double-wall connector shall be 100 percent of the height of the chimney or vent except for engineered systems. For a chimney or vent system serving multiple appliances, the maximum length of an individual connector, from the appliance outlet to the junction with the common vent or another connector, shall be 100 percent of the height of the chimney or vent.

503.10.10 Support. A vent connector shall be supported for the design and weight of the material employed to maintain clearances and prevent physical damage and separation of joints.

503.10.11 Chimney connection. Where entering a flue in a masonry or metal chimney, the vent connector shall be installed above the extreme bottom to avoid stoppage. Where a thimble or slip joint is used to facilitate removal of the connector, the connector shall be firmly attached to or inserted into the thimble or slip joint to prevent the connector from falling out. Means shall be employed to prevent the connector from entering so far as to restrict the space between its end and the opposite wall of the chimney flue (see Section 501.9).

503.10.12 Inspection. The entire length of a vent connector shall be provided with ready access for inspection, cleaning, and replacement.

503.10.13 Fireplaces. A vent connector shall not be connected to a chimney flue serving a fireplace unless the fireplace flue opening is permanently sealed.

503.10.14 Passage through ceilings, floors or walls. Single-wall metal pipe connectors shall not pass through any wall, floor or ceiling except as permitted by Sections 503.7.4 and 503.10.15.

503.10.15 Single-wall connector penetrations of combustible walls. A vent connector made of a single-wall metal pipe shall not pass through a combustible exterior wall unless guarded at the point of passage by a ventilated metal thimble not smaller than the following:

1. For listed appliances equipped with draft hoods and appliances listed for use with Type B gas vents, the thimble shall be not less than 4 inches (102 mm) larger in diameter than the vent connector. Where there is a run of not less than 6 feet (1829 mm) of vent connector in the open between the draft hood outlet and the thimble, the thimble shall be permitted to be not less than 2 inches (51 mm) larger in diameter than the vent connector.

2. For unlisted appliances having draft hoods, the thimble shall be not less than 6 inches (152 mm) larger in diameter than the vent connector.

3. For residential and low-heat appliances, the thimble shall be not less than 12 inches (305 mm) larger in diameter than the vent connector.

> **Exception:** In lieu of thimble protection, all combustible material in the wall shall be removed from the vent connector a sufficient distance to provide the specified clearance from such vent connector to combustible material. Any material used to close up such opening shall be noncombustible.

503.10.16 Medium-heat connectors. Vent connectors for medium-heat appliances shall not pass through walls or partitions constructed of combustible material.

503.11 Vent connectors for Category II, III and IV appliances. Vent connectors for Category II, III and IV appliances shall be as specified for the venting systems in accordance with Section 503.4.

503.12 Draft hoods and draft controls. The installation of draft hoods and draft controls shall comply with Sections 503.12.1 through 503.12.7.

503.12.1 Appliances requiring draft hoods. Vented appliances shall be installed with draft hoods.

Exception: Dual oven-type combination ranges; incinerators; direct-vent appliances; fan-assisted combustion system appliances; appliances requiring chimney draft for operation; single firebox boilers equipped with conversion burners with inputs greater than 400,000 Btu per hour (117 kw); appliances equipped with blast, power or pressure burners that are not listed for use with draft hoods; and appliances designed for forced venting.

503.12.2 Installation. A draft hood supplied with or forming a part of a listed vented appliance shall be installed without alteration, exactly as furnished and specified by the appliance manufacturer.

503.12.2.1 Draft hood required. If a draft hood is not supplied by the appliance manufacturer where one is required, a draft hood shall be installed, shall be of a listed or approved type and, in the absence of other instructions, shall be of the same size as the appliance flue collar. Where a draft hood is required with a conversion burner, it shall be of a listed or approved type.

503.12.2.2 Special design draft hood. Where it is determined that a draft hood of special design is needed or preferable for a particular installation, the installation shall be in accordance with the recommendations of the appliance manufacturer and shall be approved.

503.12.3 Draft control devices. Where a draft control device is part of the appliance or is supplied by the appliance manufacturer, it shall be installed in accordance with the manufacturer's instructions. In the absence of manufacturer's instructions, the device shall be attached to the flue collar of the appliance or as near to the appliance as practical.

503.12.4 Additional devices. Appliances (except incinerators) requiring a controlled chimney draft shall be permitted to be equipped with a listed double-acting barometric-draft regulator installed and adjusted in accordance with the manufacturer's instructions.

503.12.5 Location. Draft hoods and barometric draft regulators shall be installed in the same room or enclosure as the appliance in such a manner as to prevent any difference in pressure between the hood or regulator and the combustion air supply.

503.12.6 Positioning. Draft hoods and draft regulators shall be installed in the position for which they were designed with reference to the horizontal and vertical planes and shall be located so that the relief opening is not obstructed by any part of the appliance or adjacent construction. The appliance and its draft hood shall be located so that the relief opening is accessible for checking vent operation.

503.12.7 Clearance. A draft hood shall be located so its relief opening is not less than 6 inches (152 mm) from any surface except that of the appliance it serves and the venting system to which the draft hood is connected. Where a greater or lesser clearance is indicated on the appliance label, the clearance shall be not less than that specified on the label. Such clearances shall not be reduced.

503.13 Manually operated dampers. A manually operated damper shall not be placed in the vent connector for any appliance. Fixed baffles shall not be classified as manually operated dampers.

503.14 Automatically operated vent dampers. An automatically operated vent damper shall be of a listed type.

503.15 Obstructions. Devices that retard the flow of vent gases shall not be installed in a vent connector, chimney or vent. The following shall not be considered as obstructions:

1. Draft regulators and safety controls specifically listed for installation in venting systems and installed in accordance with the manufacturer's installation instructions.

2. Approved draft regulators and safety controls that are designed and installed in accordance with approved engineering methods.

3. Listed heat reclaimers and automatically operated vent dampers installed in accordance with the manufacturer's installation instructions.

4. Approved economizers, heat reclaimers and recuperators installed in venting systems of appliances not required to be equipped with draft hoods, provided that the appliance manufacturer's instructions cover the installation of such a device in the venting system and performance in accordance with Sections 503.3 and 503.3.1 is obtained.

5. Vent dampers serving listed appliances installed in accordance with Sections 504.2.1 and 504.3.1 or other approved engineering methods.

SECTION 504 (IFGS)
SIZING OF CATEGORY I APPLIANCE VENTING SYSTEMS

504.1 Definitions. The following definitions apply to the tables in this section.

APPLIANCE CATEGORIZED VENT DIAMETER/AREA. The minimum vent area/diameter permissible for Category I appliances to maintain a nonpositive vent static pressure when tested in accordance with nationally recognized standards.

FAN-ASSISTED COMBUSTION SYSTEM. An appliance equipped with an integral mechanical means to either draw or force products of combustion through the combustion chamber or heat exchanger.

FAN Min. The minimum input rating of a Category I fan-assisted appliance attached to a vent or connector.

FAN Max. The maximum input rating of a Category I fan-assisted appliance attached to a vent or connector.

NAT Max. The maximum input rating of a Category I draft-hood-equipped appliance attached to a vent or connector.

FAN + FAN. The maximum combined appliance input rating of two or more Category I fan-assisted appliances attached to the common vent.

FAN + NAT. The maximum combined appliance input rating of one or more Category I fan-assisted appliances and one or more Category I draft-hood-equipped appliances attached to the common vent.

NA. Vent configuration is not allowed due to potential for condensate formation or pressurization of the venting system, or not applicable due to physical or geometric restraints.

NAT + NAT. The maximum combined appliance input rating of two or more Category I draft-hood-equipped appliances attached to the common vent.

504.2 Application of single-appliance vent Tables 504.2(1) through 504.2(6). The application of Tables 504.2(1) through 504.2(6) shall be subject to the requirements of Sections 504.2.1 through 504.2.16.

504.2.1 Vent obstructions. These venting tables shall not be used where obstructions, as described in Section 503.15, are installed in the venting system. The installation of vents serving listed appliances with vent dampers shall be in accordance with the appliance manufacturer's instructions or in accordance with the following:

1. The maximum capacity of the vent system shall be determined using the "NAT Max" column.

2. The minimum capacity shall be determined as if the appliance were a fan-assisted appliance, using the "FAN Min" column to determine the minimum capacity of the vent system. Where the corresponding "FAN Min" is "NA," the vent configuration shall not be permitted and an alternative venting configuration shall be utilized.

504.2.2 Minimum size. Where the vent size determined from the tables is smaller than the appliance draft hood outlet or flue collar, the smaller size shall be permitted to be used provided that all of the following requirements are met:

1. The total vent height (H) is at least 10 feet (3048 mm).

2. Vents for appliance draft hood outlets or flue collars 12 inches (305 mm) in diameter or smaller are not reduced more than one table size.

3. Vents for appliance draft hood outlets or flue collars larger than 12 inches (305 mm) in diameter are not reduced more than two table sizes.

4. The maximum capacity listed in the tables for a fan-assisted appliance is reduced by 10 percent (0.90 × maximum table capacity).

5. The draft hood outlet is greater than 4 inches (102 mm) in diameter. Do not connect a 3-inch-diameter (76 mm) vent to a 4-inch-diameter (102 mm) draft hood outlet. This provision shall not apply to fan-assisted appliances.

504.2.3 Vent offsets. Single-appliance venting configurations with zero (0) lateral lengths in Tables 504.2(1), 504.2(2) and 504.2(5) shall not have elbows in the venting system. Single-appliance venting configurations with lateral lengths include two 90-degree (1.57 rad) elbows. For each additional elbow up to and including 45 degrees (0.79 rad), the maximum capacity listed in the venting tables shall be reduced by 5 percent. For each additional elbow greater than 45 degrees (0.79 rad) up to and including 90 degrees (1.57 rad), the maximum capacity listed in the venting tables shall be reduced by 10 percent.

504.2.4 Zero lateral. Zero (0) lateral (L) shall apply only to a straight vertical vent attached to a top outlet draft hood or flue collar.

504.2.5 High-altitude installations. Sea-level input ratings shall be used when determining maximum capacity for high altitude installation. Actual input (derated for altitude) shall be used for determining minimum capacity for high altitude installation.

504.2.6 Multiple input rate appliances. For appliances with more than one input rate, the minimum vent capacity (FAN Min) determined from the tables shall be less than the lowest appliance input rating, and the maximum vent capacity (FAN Max/NAT Max) determined from the tables shall be greater than the highest appliance rating input.

504.2.7 Liner system sizing and connections. Listed corrugated metallic chimney liner systems in masonry chimneys shall be sized by using Table 504.2(1) or 504.2(2) for Type B vents with the maximum capacity reduced by 20 percent (0.80 × maximum capacity) and the minimum capacity as shown in Table 504.2(1) or 504.2(2). Corrugated metallic liner systems installed with bends or offsets shall have their maximum capacity further reduced in accordance with Section 504.2.3. The 20-percent reduction for corrugated metallic chimney liner systems includes an allowance for one long-radius 90-degree (1.57 rad) turn at the bottom of the liner.

Connections between chimney liners and listed double-wall connectors shall be made with listed adapters designed for such purpose.

TABLE 504.2(1)
TYPE B DOUBLE-WALL GAS VENT

Number of Appliances	Single
Appliance Type	Category I
Appliance Vent Connection	Connected directly to vent

APPLIANCE INPUT RATING IN THOUSANDS OF BTU/H

HEIGHT (H) (feet)	LATERAL (L) (feet)	3 FAN Min	3 FAN Max	3 NAT Max	4 FAN Min	4 FAN Max	4 NAT Max	5 FAN Min	5 FAN Max	5 NAT Max	6 FAN Min	6 FAN Max	6 NAT Max	7 FAN Min	7 FAN Max	7 NAT Max	8 FAN Min	8 FAN Max	8 NAT Max	9 FAN Min	9 FAN Max	9 NAT Max
6	0	0	78	46	0	152	86	0	251	141	0	375	205	0	524	285	0	698	370	0	897	470
	2	13	51	36	18	97	67	27	157	105	32	232	157	44	321	217	53	425	285	63	543	370
	4	21	49	34	30	94	64	39	153	103	50	227	153	66	316	211	79	419	279	93	536	362
	6	25	46	32	36	91	61	47	149	100	59	223	149	78	310	205	93	413	273	110	530	354
8	0	0	84	50	0	165	94	0	276	155	0	415	235	0	583	320	0	780	415	0	1,006	537
	2	12	57	40	16	109	75	25	178	120	28	263	180	42	365	247	50	483	322	60	619	418
	5	23	53	38	32	103	71	42	171	115	53	255	173	70	356	237	83	473	313	99	607	407
	8	28	49	35	39	98	66	51	164	109	64	247	165	84	347	227	99	463	303	117	596	396
10	0	0	88	53	0	175	100	0	295	166	0	447	255	0	631	345	0	847	450	0	1,096	585
	2	12	61	42	17	118	81	23	194	129	26	289	195	40	402	273	48	533	355	57	684	457
	5	23	57	40	32	113	77	41	187	124	52	280	188	68	392	263	81	522	346	95	671	446
	10	30	51	36	41	104	70	54	176	115	67	267	175	88	376	245	104	504	330	122	651	427
15	0	0	94	58	0	191	112	0	327	187	0	502	285	0	716	390	0	970	525	0	1,263	682
	2	11	69	48	15	136	93	20	226	150	22	339	225	38	475	316	45	633	414	53	815	544
	5	22	65	45	30	130	87	39	219	142	49	330	217	64	463	300	76	620	403	90	800	529
	10	29	59	41	40	121	82	51	206	135	64	315	208	84	445	288	99	600	386	116	777	507
	15	35	53	37	48	112	76	61	195	128	76	301	198	98	429	275	115	580	373	134	755	491
20	0	0	97	61	0	202	119	0	349	202	0	540	307	0	776	430	0	1,057	575	0	1,384	752
	2	10	75	51	14	149	100	18	250	166	20	377	249	33	531	346	41	711	470	50	917	612
	5	21	71	48	29	143	96	38	242	160	47	367	241	62	519	337	73	697	460	86	902	599
	10	28	64	44	38	133	89	50	229	150	62	351	228	81	499	321	95	675	443	112	877	576
	15	34	58	40	46	124	84	59	217	142	73	337	217	94	481	308	111	654	427	129	853	557
	20	48	52	35	55	116	78	69	206	134	84	322	206	107	464	295	125	634	410	145	830	537

(continued)

TABLE 504.2(1)—continued
TYPE B DOUBLE-WALL GAS VENT

Number of Appliances	Single
Appliance Type	Category I
Appliance Vent Connection	Connected directly to vent

VENT DIAMETER—(D) inches

APPLIANCE INPUT RATING IN THOUSANDS OF BTU/H

HEIGHT (H) (feet)	LATERAL (L) (feet)	3 FAN Min	3 FAN Max	3 NAT Max	4 FAN Min	4 FAN Max	4 NAT Max	5 FAN Min	5 FAN Max	5 NAT Max	6 FAN Min	6 FAN Max	6 NAT Max	7 FAN Min	7 FAN Max	7 NAT Max	8 FAN Min	8 FAN Max	8 NAT Max	9 FAN Min	9 FAN Max	9 NAT Max
30	0	0	100	64	0	213	128	0	374	220	0	587	336	0	853	475	0	1,173	650	0	1,548	855
	2	9	81	56	13	166	112	14	283	185	18	432	280	27	613	394	33	826	535	42	1,072	700
	5	21	77	54	28	160	108	36	275	176	45	421	273	58	600	385	69	811	524	82	1,055	688
	10	27	70	50	37	150	102	48	262	171	59	405	261	77	580	371	91	788	507	107	1,028	668
	15	33	64	NA	44	141	96	57	249	163	70	389	249	90	560	357	105	765	490	124	1,002	648
	20	56	58	NA	53	132	90	66	237	154	80	374	237	102	542	343	119	743	473	139	977	628
	30	NA	NA	NA	73	113	NA	88	214	NA	104	346	219	131	507	321	149	702	444	171	929	594
50	0	0	101	67	0	216	134	0	397	232	0	633	363	0	932	518	0	1,297	708	0	1,730	952
	2	8	86	61	11	183	122	14	320	206	15	497	314	22	715	445	26	975	615	33	1,276	813
	5	20	82	NA	27	177	119	35	312	200	43	487	308	55	702	438	65	960	605	77	1,259	798
	10	26	76	NA	35	168	114	45	299	190	56	471	298	73	681	426	86	935	589	101	1,230	773
	15	59	70	NA	42	158	NA	54	287	180	66	455	288	85	662	413	100	911	572	117	1,203	747
	20	NA	NA	NA	50	149	NA	63	275	169	76	440	278	97	642	401	113	888	556	131	1,176	722
	30	NA	NA	NA	69	131	NA	84	250	NA	99	410	259	123	605	376	141	844	522	161	1,125	670
100	0	NA	NA	NA	0	218	NA	0	407	NA	0	665	400	0	997	560	0	1,411	770	0	1,908	1,040
	2	NA	NA	NA	10	194	NA	12	354	NA	13	566	375	18	831	510	21	1,155	700	25	1,536	935
	5	NA	NA	NA	26	189	NA	33	347	NA	40	557	369	52	820	504	60	1,141	692	71	1,519	926
	10	NA	NA	NA	33	182	NA	43	335	NA	53	542	361	68	801	493	80	1,118	679	94	1,492	910
	15	NA	NA	NA	40	174	NA	50	321	NA	62	528	353	80	782	482	93	1,095	666	109	1,465	895
	20	NA	NA	NA	47	166	NA	59	311	NA	71	513	344	90	763	471	105	1,073	653	122	1,438	880
	30	NA	NA	NA	NA	NA	NA	78	290	NA	92	483	NA	115	726	449	131	1,029	627	149	1,387	849
	50	NA	NA	NA	NA	NA	NA	NA	NA	NA	147	428	NA	180	651	405	197	944	575	217	1,288	787

(continued)

TABLE 504.2(1)—continued
TYPE B DOUBLE-WALL GAS VENT

Number of Appliances	Single
Appliance Type	Category I
Appliance Vent Connection	Connected directly to vent

APPLIANCE INPUT RATING IN THOUSANDS OF BTU/H

HEIGHT (H) (feet)	LATERAL (L) (feet)	10 FAN Min	10 FAN Max	10 NAT Max	12 FAN Min	12 FAN Max	12 NAT Max	14 FAN Min	14 FAN Max	14 NAT Max	16 FAN Min	16 FAN Max	16 NAT Max	18 FAN Min	18 FAN Max	18 NAT Max	20 FAN Min	20 FAN Max	20 NAT Max	22 FAN Min	22 FAN Max	22 NAT Max	24 FAN Min	24 FAN Max	24 NAT Max
6	0	0	1,121	570	0	1,645	850	0	2,267	1,170	0	2,983	1,530	0	3,802	1,960	0	4,721	2,430	0	5,737	2,950	0	6,853	3,520
6	2	75	675	455	103	982	650	138	1,346	890	178	1,769	1,170	225	2,250	1,480	296	2,782	1,850	360	3,377	2,220	426	4,030	2,670
6	4	110	668	445	147	975	640	191	1,338	880	242	1,761	1,160	300	2,242	1,475	390	2,774	1,835	469	3,370	2,215	555	4,023	2,660
6	6	128	661	435	171	967	630	219	1,330	870	276	1,753	1,150	341	2,235	1,470	437	2,767	1,820	523	3,363	2,210	618	4,017	2,650
8	0	0	1,261	660	0	1,858	970	0	2,571	1,320	0	3,399	1,740	0	4,333	2,220	0	5,387	2,750	0	6,555	3,360	0	7,838	4,010
8	2	71	770	515	98	1,124	745	130	1,543	1,020	168	2,030	1,340	212	2,584	1,700	278	3,196	2,110	336	3,882	2,560	401	4,634	3,050
8	5	115	758	503	154	1,110	733	199	1,528	1,010	251	2,013	1,330	311	2,563	1,685	398	3,180	2,090	476	3,863	2,545	562	4,612	3,040
8	8	137	746	490	180	1,097	720	231	1,514	1,000	289	2,000	1,320	354	2,552	1,670	450	3,163	2,070	537	3,850	2,530	630	4,602	3,030
10	0	0	1,377	720	0	2,036	1,060	0	2,825	1,450	0	3,742	1,925	0	4,782	2,450	0	5,955	3,050	0	7,254	3,710	0	8,682	4,450
10	2	68	852	560	93	1,244	850	124	1,713	1,130	161	2,256	1,480	202	2,868	1,890	264	3,556	2,340	319	4,322	2,840	378	5,153	3,390
10	5	112	839	547	149	1,229	829	192	1,696	1,105	243	2,238	1,461	300	2,849	1,871	382	3,536	2,318	458	4,301	2,818	540	5,132	3,371
10	10	142	817	525	187	1,204	795	238	1,669	1,080	298	2,209	1,430	364	2,818	1,840	459	3,504	2,280	546	4,268	2,780	641	5,099	3,340
15	0	0	1,596	840	0	2,380	1,240	0	3,323	1,720	0	4,423	2,270	0	5,678	2,900	0	7,099	3,620	0	8,665	4,410	0	10,393	5,300
15	2	63	1,019	675	86	1,495	985	114	2,062	1,350	147	2,719	1,770	186	3,467	2,260	239	4,304	2,800	290	5,232	3,410	346	6,251	4,080
15	5	105	1,003	660	140	1,476	967	182	2,041	1,327	229	2,696	1,748	283	3,442	2,235	355	4,278	2,777	426	5,204	3,385	501	6,222	4,057
15	10	135	977	635	177	1,446	936	227	2,009	1,289	283	2,659	1,712	346	3,402	2,193	432	4,234	2,739	510	5,159	3,343	599	6,175	4,019
15	15	155	953	610	202	1,418	905	257	1,976	1,250	318	2,623	1,675	385	3,363	2,150	479	4,192	2,700	564	5,115	3,300	665	6,129	3,980
20	0	0	1,756	930	0	2,637	1,350	0	3,701	1,900	0	4,948	2,520	0	6,376	3,250	0	7,988	4,060	0	9,785	4,980	0	11,753	6,000
20	2	59	1,150	755	81	1,694	1,100	107	2,343	1,520	139	3,097	2,000	175	3,955	2,570	220	4,916	3,200	269	5,983	3,910	321	7,154	4,700
20	5	101	1,133	738	135	1,674	1,079	174	2,320	1,498	219	3,071	1,978	270	3,926	2,544	337	4,885	3,174	403	5,950	3,880	475	7,119	4,662
20	10	130	1,105	710	172	1,641	1,045	220	2,282	1,460	273	3,029	1,940	334	3,880	2,500	413	4,835	3,130	489	5,896	3,830	573	7,063	4,600
20	15	150	1,078	688	195	1,609	1,018	248	2,245	1,425	306	2,988	1,910	372	3,835	2,465	459	4,786	3,090	541	5,844	3,795	631	7,007	4,575
20	20	167	1,052	665	217	1,578	990	273	2,210	1,390	335	2,948	1,880	404	3,791	2,430	495	4,737	3,050	585	5,792	3,760	689	6,953	4,550

VENT DIAMETER—(D) inches

(continued)

TABLE 504.2(1)—continued
TYPE B DOUBLE-WALL GAS VENT

Number of Appliances	Single
Appliance Type	Category I
Appliance Vent Connection	Connected directly to vent

VENT DIAMETER—(D) inches

APPLIANCE INPUT RATING IN THOUSANDS OF BTU/H

HEIGHT (H) (feet)	LATERAL (L) (feet)	10 FAN Min	10 FAN Max	10 NAT Max	12 FAN Min	12 FAN Max	12 NAT Max	14 FAN Min	14 FAN Max	14 NAT Max	16 FAN Min	16 FAN Max	16 NAT Max	18 FAN Min	18 FAN Max	18 NAT Max	20 FAN Min	20 FAN Max	20 NAT Max	22 FAN Min	22 FAN Max	22 NAT Max	24 FAN Min	24 FAN Max	24 NAT Max
30	0	0	1,977	1,060	0	3,004	1,550	0	4,252	2,170	0	5,725	2,920	0	7,420	3,770	0	9,341	4,750	0	11,483	5,850	0	13,848	7,060
	2	54	1,351	865	74	2,004	1,310	98	2,786	1,800	127	3,696	2,380	159	4,734	3,050	199	5,900	3,810	241	7,194	4,650	285	8,617	5,600
	5	96	1,332	851	127	1,981	1,289	164	2,759	1,775	206	3,666	2,350	252	4,701	3,020	312	5,863	3,783	373	7,155	4,622	439	8,574	5,552
	10	125	1,301	829	164	1,944	1,254	209	2,716	1,733	259	3,617	2,300	316	4,647	2,970	386	5,803	3,739	456	7,090	4,574	535	8,505	5,471
	15	143	1,272	807	187	1,908	1,220	237	2,674	1,692	292	3,570	2,250	354	4,594	2,920	431	5,744	3,695	507	7,026	4,527	590	8,437	5,391
	20	160	1,243	784	207	1,873	1,185	260	2,633	1,650	319	3,523	2,200	384	4,542	2,870	467	5,686	3,650	548	6,964	4,480	639	8,370	5,310
	30	195	1,189	745	246	1,807	1,130	305	2,555	1,585	369	3,433	2,130	440	4,442	2,785	540	5,574	3,565	635	6,842	4,375	739	8,239	5,225
50	0	0	2,231	1,195	0	3,441	1,825	0	4,934	2,550	0	6,711	3,440	0	8,774	4,460	0	11,129	5,635	0	13,767	6,940	0	16,694	8,430
	2	41	1,620	1,010	66	2,431	1,513	86	3,409	2,125	113	4,554	2,840	141	5,864	3,670	171	7,339	4,630	209	8,980	5,695	251	10,788	6,860
	5	90	1,600	996	118	2,406	1,495	151	3,380	2,102	191	4,520	2,813	234	5,826	3,639	283	7,295	4,597	336	8,933	5,654	394	10,737	6,818
	10	118	1,567	972	154	2,366	1,466	196	3,332	2,064	243	4,464	2,767	295	5,763	3,585	355	7,224	4,542	419	8,855	5,585	491	10,652	6,749
	15	136	1,536	948	177	2,327	1,437	222	3,285	2,026	274	4,409	2,721	330	5,701	3,534	396	7,155	4,511	465	8,779	5,546	542	10,570	6,710
	20	151	1,505	924	195	2,288	1,408	244	3,239	1,987	300	4,356	2,675	361	5,641	3,481	433	7,086	4,479	506	8,704	5,506	586	10,488	6,670
	30	183	1,446	876	232	2,214	1,349	287	3,150	1,910	347	4,253	2,631	412	5,523	3,431	494	6,953	4,421	577	8,557	5,444	672	10,328	6,603
100	0	0	2,491	1,310	0	3,925	2,050	0	5,729	2,950	0	7,914	4,050	0	10,485	5,300	0	13,454	6,700	0	16,817	8,600	0	20,578	10,300
	2	30	1,975	1,170	44	3,027	1,820	72	4,313	2,550	95	5,834	3,500	120	7,591	4,600	138	9,577	5,800	169	11,803	7,200	204	14,264	8,800
	5	82	1,955	1,159	107	3,002	1,803	136	4,282	2,531	172	5,797	3,475	208	7,548	4,566	245	9,528	5,769	293	11,748	7,162	341	14,204	8,756
	10	108	1,923	1,142	142	2,961	1,775	180	4,231	2,500	223	5,737	3,434	268	7,478	4,509	318	9,447	5,717	374	11,658	7,100	436	14,105	8,683
	15	126	1,892	1,124	163	2,920	1,747	206	4,182	2,469	252	5,678	3,392	304	7,409	4,451	358	9,367	5,665	418	11,569	7,037	487	14,007	8,610
	20	141	1,861	1,107	181	2,880	1,719	226	4,133	2,438	277	5,619	3,351	330	7,341	4,394	387	9,289	5,613	452	11,482	6,975	523	13,910	8,537
	30	170	1,802	1,071	215	2,803	1,663	265	4,037	2,375	319	5,505	3,267	378	7,209	4,279	446	9,136	5,509	514	11,310	6,850	592	13,720	8,391
	50	241	1,688	1,000	292	2,657	1,550	350	3,856	2,250	415	5,289	3,100	486	6,956	4,050	572	8,841	5,300	659	10,979	6,600	752	13,354	8,100

For SI: 1 inch = 25.4 mm, 1 foot = 304.8 mm, 1 British thermal unit per hour = 0.2931 W.

TABLE 504.2(2)
TYPE B DOUBLE-WALL GAS VENT

Number of Appliances	Single
Appliance Type	Category I
Appliance Vent Connection	Single-wall metal connector

VENT DIAMETER—(D) inches
APPLIANCE INPUT RATING IN THOUSANDS OF BTU/H

HEIGHT (H) (feet)	LATERAL (L) (feet)	3 FAN Min	3 FAN Max	3 NAT Max	4 FAN Min	4 FAN Max	4 NAT Max	5 FAN Min	5 FAN Max	5 NAT Max	6 FAN Min	6 FAN Max	6 NAT Max	7 FAN Min	7 FAN Max	7 NAT Max	8 FAN Min	8 FAN Max	8 NAT Max	9 FAN Min	9 FAN Max	9 NAT Max	10 FAN Min	10 FAN Max	10 NAT Max	12 FAN Min	12 FAN Max	12 NAT Max
6	0	38	77	45	59	151	85	85	249	140	126	373	204	165	522	284	211	695	369	267	894	469	371	1,118	569	537	1,639	849
	2	39	51	36	60	96	66	85	156	104	123	231	156	159	320	213	201	423	284	251	541	368	347	673	453	498	979	648
	4	NA	NA	33	74	92	63	102	152	102	146	225	152	187	313	208	237	416	277	295	533	360	409	664	443	584	971	638
	6	NA	NA	31	83	89	60	114	147	99	163	220	148	207	307	203	263	409	271	327	526	352	449	656	433	638	962	627
8	0	37	83	50	58	164	93	83	273	154	123	412	234	161	580	319	206	777	414	258	1,002	536	360	1,257	658	521	1,852	967
	2	39	56	39	59	108	75	83	176	119	121	261	179	155	363	246	197	482	321	246	617	417	339	768	513	486	1,120	743
	5	NA	NA	37	77	102	69	107	168	114	151	252	171	193	352	235	245	470	311	305	604	404	418	754	500	598	1,104	730
	8	NA	NA	33	90	95	64	122	161	107	175	243	163	223	342	225	280	458	300	344	591	392	470	740	486	665	1,089	715
10	0	37	87	53	57	174	99	82	293	165	120	444	254	158	628	344	202	844	449	253	1,093	584	351	1,373	718	507	2,031	1,057
	2	39	61	41	59	117	80	82	193	128	119	287	194	153	400	272	193	531	354	242	681	456	332	849	559	475	1,242	848
	5	52	56	39	76	111	76	105	185	122	148	277	186	190	388	261	241	518	344	299	667	443	409	834	544	584	1,224	825
	10	NA	NA	34	97	100	68	132	171	112	188	261	171	237	369	241	296	497	325	363	643	423	492	808	520	688	1,194	788
15	0	36	93	57	56	190	111	80	325	186	116	499	283	153	713	388	195	966	523	244	1,259	681	336	1,591	838	488	2,374	1,237
	2	38	69	47	57	136	93	80	225	149	115	337	224	148	473	314	187	631	413	232	812	543	319	1,015	673	457	1,491	983
	5	51	63	44	75	128	86	102	216	140	144	326	217	182	459	298	231	616	400	287	795	526	392	997	657	562	1,469	963
	10	NA	NA	39	95	116	79	128	201	131	182	308	203	228	438	284	284	592	381	349	768	501	470	966	628	664	1,433	928
	15	NA	NA	NA	NA	NA	72	158	186	124	220	290	192	272	418	269	334	568	367	404	742	484	540	937	601	750	1,399	894
20	0	35	96	60	54	200	118	78	346	201	114	537	306	149	772	428	190	1,053	573	238	1,379	750	326	1,751	927	473	2,631	1,346
	2	37	74	50	56	148	99	78	248	165	113	375	248	144	528	344	182	708	468	227	914	611	309	1,146	754	443	1,689	1,098
	5	50	68	47	73	140	94	100	239	158	141	363	239	178	514	334	224	692	457	279	896	596	381	1,126	734	547	1,665	1,074
	10	NA	NA	41	93	129	86	125	223	146	177	344	224	222	491	316	277	666	437	339	866	570	457	1,092	702	646	1,626	1,037
	15	NA	NA	NA	NA	NA	80	155	208	136	216	325	210	264	469	301	325	640	419	393	838	549	526	1,060	677	730	1,587	1,005
	20	NA	NA	NA	NA	NA	NA	186	192	126	254	306	196	309	448	285	374	616	400	448	810	526	592	1,028	651	808	1,550	973

(continued)

TABLE 504.2(2)—continued
TYPE B DOUBLE-WALL GAS VENT

Number of Appliances	Single
Appliance Type	Category I
Appliance Vent Connection	Single-wall metal connector

APPLIANCE INPUT RATING IN THOUSANDS OF BTU/H

HEIGHT (H) (feet)	LATERAL (L) (feet)	3 FAN Min	3 FAN Max	3 NAT Max	4 FAN Min	4 FAN Max	4 NAT Max	5 FAN Min	5 FAN Max	5 NAT Max	6 FAN Min	6 FAN Max	6 NAT Max	7 FAN Min	7 FAN Max	7 NAT Max	8 FAN Min	8 FAN Max	8 NAT Max	9 FAN Min	9 FAN Max	9 NAT Max	10 FAN Min	10 FAN Max	10 NAT Max	12 FAN Min	12 FAN Max	12 NAT Max
30	0	34	99	63	53	211	127	76	372	219	110	584	334	144	849	472	184	1,168	647	229	1,542	852	312	1,971	1,056	454	2,996	1,545
	2	37	80	56	55	164	111	76	281	183	109	429	279	139	610	392	175	823	533	219	1,069	698	296	1,346	863	424	1,999	1,308
	5	49	74	52	72	157	106	98	271	173	136	417	271	171	595	382	215	806	521	269	1,049	684	366	1,324	846	524	1,971	1,283
	10	NA	NA	NA	91	144	98	122	255	168	171	397	257	213	570	367	265	777	501	327	1,017	662	440	1,287	821	620	1,927	1,234
	15	NA	NA	NA	115	131	NA	151	239	157	208	377	242	255	547	349	312	750	481	379	985	638	507	1,251	794	702	1,884	1,205
	20	NA	NA	NA	NA	NA	NA	181	223	NA	246	357	228	298	524	333	360	723	461	433	955	615	570	1,216	768	780	1,841	1,166
	30	NA	NA	NA	NA	NA	NA	NA	NA	NA	NA	NA	NA	389	477	305	461	670	426	541	895	574	704	1,147	720	937	1,759	1,101
50	0	33	99	66	51	213	133	73	394	230	105	629	361	138	928	515	176	1,292	704	220	1,724	948	295	2,223	1,189	428	3,432	1,818
	2	36	84	61	53	181	121	73	318	205	104	495	312	133	712	443	168	971	613	209	1,273	811	280	1,615	1,007	401	2,426	1,509
	5	48	80	NA	70	174	117	94	308	198	131	482	305	164	696	435	204	953	602	257	1,252	795	347	1,591	991	496	2,396	1,490
	10	NA	NA	NA	89	160	NA	118	292	186	162	461	292	203	671	420	253	923	583	313	1,217	765	418	1,551	963	589	2,347	1,455
	15	NA	NA	NA	112	148	NA	145	275	174	199	441	280	244	646	405	299	894	562	363	1,183	736	481	1,512	934	668	2,299	1,421
	20	NA	NA	NA	NA	NA	NA	176	257	NA	236	420	267	285	622	389	345	866	543	415	1,150	708	544	1,473	906	741	2,251	1,387
	30	NA	NA	NA	NA	NA	NA	NA	NA	NA	315	376	NA	373	573	NA	442	809	502	521	1,086	649	674	1,399	848	892	2,159	1,318
100	0	NA	NA	NA	49	214	NA	69	403	NA	100	659	395	131	991	555	166	1,404	765	207	1,900	1,033	273	2,479	1,300	395	3,912	2,042
	2	NA	NA	NA	51	192	NA	70	351	NA	98	563	373	125	828	508	158	1,152	698	196	1,532	933	259	1,970	1,168	371	3,021	1,817
	5	NA	NA	NA	67	186	NA	90	342	NA	125	551	366	156	813	501	194	1,134	688	240	1,511	921	322	1,945	1,153	460	2,990	1,796
	10	NA	NA	NA	85	175	NA	113	324	NA	153	532	354	191	789	486	238	1,104	672	293	1,477	902	389	1,905	1,133	547	2,938	1,763
	15	NA	NA	NA	132	162	NA	138	310	NA	188	511	343	230	764	473	281	1,075	656	342	1,443	884	447	1,865	1,110	618	2,888	1,730
	20	NA	NA	NA	NA	NA	NA	168	295	NA	224	487	NA	270	739	458	325	1,046	639	391	1,410	864	507	1,825	1,087	690	2,838	1,696
	30	NA	NA	NA	NA	NA	NA	231	264	NA	301	448	NA	355	685	NA	418	988	NA	491	1,343	824	631	1,747	1,041	834	2,739	1,627
	50	NA	NA	NA	NA	NA	NA	NA	NA	NA	NA	NA	NA	540	584	NA	617	866	NA	711	1,205	NA	895	1,591	NA	1,138	2,547	1,489

For SI: 1 inch = 25.4 mm, 1 foot = 304.8 mm, 1 British thermal unit per hour = 0.2931 W.

TABLE 504.2(3)
MASONRY CHIMNEY

Number of Appliances	Single
Appliance Type	Category I
Appliance Vent Connection	Type B double-wall connector

TYPE B DOUBLE-WALL CONNECTOR DIAMETER—(D) inches
to be used with chimney areas within the size limits at bottom

APPLIANCE INPUT RATING IN THOUSANDS OF BTU/H

HEIGHT (H) (feet)	LATERAL (L) (feet)	3 FAN Min	3 FAN Max	3 NAT Max	4 FAN Min	4 FAN Max	4 NAT Max	5 FAN Min	5 FAN Max	5 NAT Max	6 FAN Min	6 FAN Max	6 NAT Max	7 FAN Min	7 FAN Max	7 NAT Max	8 FAN Min	8 FAN Max	8 NAT Max	9 FAN Min	9 FAN Max	9 NAT Max	10 FAN Min	10 FAN Max	10 NAT Max	12 FAN Min	12 FAN Max	12 NAT Max
6	2	NA	NA	28	NA	NA	52	NA	NA	86	NA	NA	130	NA	NA	180	NA	NA	247	NA	NA	320	NA	NA	401	NA	NA	581
	5	NA	NA	25	NA	NA	49	NA	NA	82	NA	NA	117	NA	NA	165	NA	NA	231	NA	NA	298	NA	NA	376	NA	NA	561
8	2	NA	NA	29	NA	NA	55	NA	NA	93	NA	NA	145	NA	NA	198	NA	NA	266	84	590	350	100	728	446	139	1,024	651
	5	NA	NA	26	NA	NA	52	NA	NA	88	NA	NA	134	NA	NA	183	NA	NA	247	NA	NA	328	149	711	423	201	1,007	640
	8	NA	NA	24	NA	NA	48	NA	NA	83	NA	NA	127	NA	NA	175	NA	NA	239	NA	NA	318	173	695	410	231	990	623
10	2	NA	NA	31	NA	NA	61	NA	NA	103	NA	NA	162	NA	NA	221	68	519	298	82	655	388	98	810	491	136	1,144	724
	5	NA	NA	28	NA	NA	57	NA	NA	96	NA	NA	148	NA	NA	204	NA	NA	277	124	638	365	146	791	466	196	1,124	712
	10	NA	NA	25	NA	NA	50	NA	NA	87	NA	NA	139	NA	NA	191	NA	NA	263	155	610	347	182	762	444	240	1,093	668
15	2	NA	NA	35	NA	NA	67	NA	NA	114	NA	NA	179	53	475	250	64	613	336	77	779	441	92	968	562	127	1,376	841
	5	NA	NA	35	NA	NA	62	NA	NA	107	NA	NA	164	NA	NA	231	99	594	313	118	759	416	139	946	533	186	1,352	828
	10	NA	NA	28	NA	NA	55	NA	NA	97	NA	NA	153	NA	NA	216	126	565	296	148	727	394	173	912	507	229	1,315	777
	15	NA	NA	NA	NA	NA	48	NA	NA	89	NA	NA	141	NA	NA	201	NA	NA	281	171	698	375	198	880	485	259	1,280	742
20	2	NA	NA	38	NA	NA	74	NA	NA	124	NA	NA	201	51	522	274	61	678	375	73	867	491	87	1,083	627	121	1,548	953
	5	NA	NA	36	NA	NA	68	NA	NA	116	NA	NA	184	80	503	254	95	658	350	113	845	463	133	1,059	597	179	1,523	933
	10	NA	NA	NA	NA	NA	60	NA	NA	107	NA	NA	172	NA	NA	237	122	627	332	143	811	440	167	1,022	566	221	1,482	879
	15	NA	NA	NA	NA	NA	NA	NA	NA	97	NA	NA	159	NA	NA	220	NA	NA	314	165	780	418	191	987	541	251	1,443	840
	20	NA	NA	NA	NA	NA	NA	NA	NA	83	NA	NA	148	NA	NA	206	NA	NA	296	186	750	397	214	955	513	277	1,406	807

(continued)

TABLE 504.2(3)—continued
MASONRY CHIMNEY

		Number of Appliances	Single
		Appliance Type	Category I
		Appliance Vent Connection	Type B double-wall connector

TYPE B DOUBLE-WALL CONNECTOR DIAMETER—(D) inches
to be used with chimney areas within the size limits at bottom

APPLIANCE INPUT RATING IN THOUSANDS OF BTU/H

HEIGHT (H) (feet)	LATERAL (L) (feet)	3 FAN Min	3 FAN Max	3 NAT Max	4 FAN Min	4 FAN Max	4 NAT Max	5 FAN Min	5 FAN Max	5 NAT Max	6 FAN Min	6 FAN Max	6 NAT Max	7 FAN Min	7 FAN Max	7 NAT Max	8 FAN Min	8 FAN Max	8 NAT Max	9 FAN Min	9 FAN Max	9 NAT Max	10 FAN Min	10 FAN Max	10 NAT Max	12 FAN Min	12 FAN Max	12 NAT Max
30	2	NA	NA	41	NA	NA	82	NA	NA	137	NA	NA	216	47	581	303	57	762	421	68	985	558	81	1,240	717	111	1,793	1,112
	5	NA	NA	NA	NA	NA	76	NA	NA	128	NA	NA	198	75	561	281	90	741	393	106	962	526	125	1,216	683	169	1,766	1,094
	10	NA	NA	NA	NA	NA	67	NA	NA	115	NA	NA	184	NA	NA	263	115	709	373	135	927	500	158	1,176	648	210	1,721	1,025
	15	NA	NA	NA	NA	NA	NA	NA	NA	107	NA	NA	171	NA	NA	243	NA	NA	353	156	893	476	181	1,139	621	239	1,679	981
	20	NA	NA	NA	NA	NA	NA	NA	NA	91	NA	NA	159	NA	NA	227	NA	NA	332	176	860	450	203	1,103	592	264	1,638	940
	30	NA	NA	NA	NA	NA	NA	NA	NA	NA	NA	NA	NA	NA	NA	188	NA	NA	288	NA	NA	416	249	1,035	555	318	1,560	877
50	2	NA	NA	NA	NA	NA	92	NA	NA	161	NA	NA	251	NA	NA	351	51	840	477	61	1,106	633	72	1,413	812	99	2,080	1,243
	5	NA	NA	NA	NA	NA	NA	NA	NA	151	NA	NA	230	NA	NA	323	83	819	445	98	1,083	596	116	1,387	774	155	2,052	1,225
	10	NA	NA	NA	NA	NA	NA	NA	NA	138	NA	NA	215	NA	NA	304	NA	NA	424	126	1,047	567	147	1,347	733	195	2,006	1,147
	15	NA	NA	NA	NA	NA	NA	NA	NA	127	NA	NA	199	NA	NA	282	NA	NA	400	146	1,010	539	170	1,307	702	222	1,961	1,099
	20	NA	NA	NA	NA	NA	NA	NA	NA	NA	NA	NA	185	NA	NA	264	NA	NA	376	165	977	511	190	1,269	669	246	1,916	1,050
	30	NA	NA	NA	NA	NA	NA	NA	NA	NA	NA	NA	NA	NA	NA	NA	NA	NA	327	NA	NA	468	233	1,196	623	295	1,832	984
Minimum Internal Area of Chimney (square inches)			12			19			28			38			50			63			78			95			132	
Maximum Internal Area of Chimney (square inches)			49			88			137			198			269			352			445			550			792	

For SI: 1 inch = 25.4 mm, 1 square inch = 645.16 mm², 1 foot = 304.8 mm, 1 British thermal unit per hour = 0.2931 W.

TABLE 504.2(4)
MASONRY CHIMNEY

Number of Appliances	Single
Appliance Type	Category I
Appliance Vent Connection	Single-wall metal connector

SINGLE-WALL METAL CONNECTOR DIAMETER—(D) inches
to be used with chimney areas within the size limits at bottom

APPLIANCE INPUT RATING IN THOUSANDS OF BTU/H

HEIGHT (H) (feet)	LATERAL (L) (feet)	3 FAN Min	3 FAN Max	3 NAT Max	4 FAN Min	4 FAN Max	4 NAT Max	5 FAN Min	5 FAN Max	5 NAT Max	6 FAN Min	6 FAN Max	6 NAT Max	7 FAN Min	7 FAN Max	7 NAT Max	8 FAN Min	8 FAN Max	8 NAT Max	9 FAN Min	9 FAN Max	9 NAT Max	10 FAN Min	10 FAN Max	10 NAT Max	12 FAN Min	12 FAN Max	12 NAT Max
6	2	NA	NA	28	NA	NA	52	NA	NA	86	NA	NA	130	NA	NA	180	NA	NA	247	NA	NA	319	NA	NA	400	NA	NA	580
6	5	NA	NA	25	NA	NA	48	NA	NA	81	NA	NA	116	NA	NA	164	NA	NA	230	NA	NA	297	NA	NA	375	NA	NA	560
8	2	NA	NA	29	NA	NA	55	NA	NA	93	NA	NA	145	NA	NA	197	NA	NA	265	NA	NA	349	382	725	445	549	1,021	650
8	5	NA	NA	26	NA	NA	51	NA	NA	87	NA	NA	133	NA	NA	182	NA	NA	246	NA	NA	327	NA	NA	422	673	1,003	638
8	8	NA	NA	23	NA	NA	47	NA	NA	82	NA	NA	126	NA	NA	174	NA	NA	237	NA	NA	317	NA	NA	408	747	985	621
10	2	NA	NA	31	NA	NA	61	NA	NA	102	NA	NA	161	NA	NA	220	216	518	297	271	654	387	373	808	490	536	1,142	722
10	5	NA	NA	28	NA	NA	56	NA	NA	95	NA	NA	147	NA	NA	203	NA	NA	276	334	635	364	459	789	465	657	1,121	710
10	10	NA	NA	24	NA	NA	49	NA	NA	86	NA	NA	137	NA	NA	189	NA	NA	261	NA	NA	345	547	758	441	771	1,088	665
15	2	NA	NA	35	NA	NA	67	NA	NA	113	NA	NA	178	166	473	249	211	611	335	264	776	440	362	965	560	520	1,373	840
15	5	NA	NA	32	NA	NA	61	NA	NA	106	NA	NA	163	NA	NA	230	261	591	312	325	775	414	444	942	531	637	1,348	825
15	10	NA	NA	27	NA	NA	54	NA	NA	96	NA	NA	151	NA	NA	214	NA	NA	294	392	722	392	531	907	504	749	1,309	774
15	15	NA	NA	NA	NA	NA	46	NA	NA	87	NA	NA	138	NA	NA	198	NA	NA	278	452	692	372	606	873	481	841	1,272	738
20	2	NA	NA	38	NA	NA	73	NA	NA	123	NA	NA	200	163	520	273	206	675	374	258	864	490	252	1,079	625	508	1,544	950
20	5	NA	NA	35	NA	NA	67	NA	NA	115	NA	NA	183	80	NA	252	255	655	348	317	842	461	433	1,055	594	623	1,518	930
20	10	NA	NA	NA	NA	NA	59	NA	NA	105	NA	NA	170	NA	NA	235	312	622	330	382	806	437	517	1,016	562	733	1,475	875
20	15	NA	NA	NA	NA	NA	NA	NA	NA	95	NA	NA	156	NA	NA	217	NA	NA	311	442	773	414	591	979	539	823	1,434	835
20	20	NA	NA	NA	NA	NA	NA	NA	NA	80	NA	NA	144	NA	NA	202	NA	NA	292	NA	NA	392	663	944	510	911	1,394	800

(continued)

TABLE 504.2(4)—continued
MASONRY CHIMNEY

Number of Appliances	Single
Appliance Type	Category I
Appliance Vent Connection	Single-wall metal connector

SINGLE-WALL METAL CONNECTOR DIAMETER—(D) inches
to be used with chimney areas within the size limits at bottom

APPLIANCE INPUT RATING IN THOUSANDS OF BTU/H

HEIGHT (H) (feet)	LATERAL (L) (feet)	3 FAN Min	3 FAN Max	3 NAT Max	4 FAN Min	4 FAN Max	4 NAT Max	5 FAN Min	5 FAN Max	5 NAT Max	6 FAN Min	6 FAN Max	6 NAT Max	7 FAN Min	7 FAN Max	7 NAT Max	8 FAN Min	8 FAN Max	8 NAT Max	9 FAN Min	9 FAN Max	9 NAT Max	10 FAN Min	10 FAN Max	10 NAT Max	12 FAN Min	12 FAN Max	12 NAT Max
30	2	NA	NA	41	NA	NA	81	NA	NA	136	NA	NA	215	158	578	302	200	759	420	249	982	556	340	1,237	715	489	1,789	1,110
	5	NA	NA	NA	NA	NA	75	NA	NA	127	NA	NA	196	NA	NA	279	245	737	391	306	958	524	417	1,210	680	600	1,760	1,090
	10	NA	NA	NA	NA	NA	66	NA	NA	113	NA	NA	182	NA	NA	260	300	703	370	370	920	496	500	1,168	644	708	1,713	1,020
	15	NA	NA	NA	NA	NA	NA	NA	NA	105	NA	NA	168	NA	NA	240	NA	NA	349	428	884	471	572	1,128	615	798	1,668	975
	20	NA	NA	NA	NA	NA	NA	NA	NA	88	NA	NA	155	NA	NA	223	NA	NA	327	NA	NA	445	643	1,089	585	883	1,624	932
	30	NA	NA	NA	NA	NA	NA	NA	NA	NA	NA	NA	NA	NA	NA	182	NA	NA	281	NA	NA	408	NA	NA	544	1,055	1,539	865
50	2	NA	NA	48	NA	NA	91	NA	NA	160	NA	NA	250	NA	NA	350	191	837	475	238	1,103	631	323	1,408	810	463	2,076	1,240
	5	NA	NA	NA	NA	NA	NA	NA	NA	149	NA	NA	228	NA	NA	321	NA	NA	442	293	1,078	593	398	1,381	770	571	2,044	1,220
	10	NA	NA	NA	NA	NA	NA	NA	NA	136	NA	NA	212	NA	NA	301	NA	NA	420	355	1,038	562	447	1,337	728	674	1,994	1,140
	15	NA	NA	NA	NA	NA	NA	NA	NA	124	NA	NA	195	NA	NA	278	NA	NA	395	NA	NA	533	546	1,294	695	761	1,945	1,090
	20	NA	NA	NA	NA	NA	NA	NA	NA	NA	NA	NA	180	NA	NA	258	NA	NA	370	NA	NA	504	616	1,251	660	844	1,898	1,040
	30	NA	NA	NA	NA	NA	NA	NA	NA	NA	NA	NA	NA	NA	NA	218	NA	NA	318	NA	NA	458	NA	NA	610	1,009	1,805	970
Minimum Internal Area of Chimney (square inches)			12			19			28			38			50			63			78			95			132	
Maximum Internal Area of Chimney (square inches)			49			88			137			198			269			352			445			550			792	

For SI: 1 inch = 25.4 mm, 1 square inch = 645.16 mm², 1 foot = 304.8 mm, 1 British thermal unit per hour = 0.2931 W.

TABLE 504.2(5)
SINGLE-WALL METAL PIPE OR TYPE B
ASBESTOS CEMENT VENT

Number of Appliances	Single
Appliance Type	Draft hood equipped
Appliance Vent Connection	Connected directly to pipe or vent

HEIGHT (H) (feet)	LATERAL (L) (feet)	VENT DIAMETER—(D) inches							
		3	4	5	6	7	8	10	12
		MAXIMUM APPLIANCE INPUT RATING IN THOUSANDS OF BTU/H							
6	0	39	70	116	170	232	312	500	750
	2	31	55	94	141	194	260	415	620
	5	28	51	88	128	177	242	390	600
8	0	42	76	126	185	252	340	542	815
	2	32	61	102	154	210	284	451	680
	5	29	56	95	141	194	264	430	648
	10	24	49	86	131	180	250	406	625
10	0	45	84	138	202	279	372	606	912
	2	35	67	111	168	233	311	505	760
	5	32	61	104	153	215	289	480	724
	10	27	54	94	143	200	274	455	700
	15	NA	46	84	130	186	258	432	666
15	0	49	91	151	223	312	420	684	1,040
	2	39	72	122	186	260	350	570	865
	5	35	67	110	170	240	325	540	825
	10	30	58	103	158	223	308	514	795
	15	NA	50	93	144	207	291	488	760
	20	NA	NA	82	132	195	273	466	726
20	0	53	101	163	252	342	470	770	1,190
	2	42	80	136	210	286	392	641	990
	5	38	74	123	192	264	364	610	945
	10	32	65	115	178	246	345	571	910
	15	NA	55	104	163	228	326	550	870
	20	NA	NA	91	149	214	306	525	832
30	0	56	108	183	276	384	529	878	1,370
	2	44	84	148	230	320	441	730	1,140
	5	NA	78	137	210	296	410	694	1,080
	10	NA	68	125	196	274	388	656	1,050
	15	NA	NA	113	177	258	366	625	1,000
	20	NA	NA	99	163	240	344	596	960
	30	NA	NA	NA	NA	192	295	540	890
50	0	NA	120	210	310	443	590	980	1,550
	2	NA	95	171	260	370	492	820	1,290
	5	NA	NA	159	234	342	474	780	1,230
	10	NA	NA	146	221	318	456	730	1,190
	15	NA	NA	NA	200	292	407	705	1,130
	20	NA	NA	NA	185	276	384	670	1,080
	30	NA	NA	NA	NA	222	330	605	1,010

For SI: 1 inch = 25.4 mm, 1 foot = 304.8 mm, 1 British thermal unit per hour = 0.2931 W.

Number of Appliances	Single
Appliance Type	NAT
Appliance Vent Connection	Type B double-wall connector

TABLE 504.2(6)
EXTERIOR MASONRY CHIMNEY

VENT HEIGHT (feet)	MINIMUM ALLOWABLE INPUT RATING OF SPACE-HEATING APPLIANCE IN THOUSANDS OF BTU PER HOUR							
	Internal area of chimney (square inches)							
	12	19	28	38	50	63	78	113
37°F or Greater	Local 99% Winter Design Temperature: 37°F or Greater							
6	0	0	0	0	0	0	0	0
8	0	0	0	0	0	0	0	0
10	0	0	0	0	0	0	0	0
15	NA	0	0	0	0	0	0	0
20	NA	NA	123	190	249	184	0	0
30	NA	NA	NA	NA	NA	393	334	0
50	NA	NA	NA	NA	NA	NA	NA	579
27 to 36°F	Local 99% Winter Design Temperature: 27 to 36°F							
6	0	0	68	116	156	180	212	266
8	0	0	82	127	167	187	214	263
10	0	51	97	141	183	201	225	265
15	NA	NA	NA	NA	233	253	274	305
20	NA	NA	NA	NA	NA	307	330	362
30	NA	NA	NA	NA	NA	419	445	485
50	NA	NA	NA	NA	NA	NA	NA	763
17 to 26°F	Local 99% Winter Design Temperature: 17 to 26°F							
6	NA	NA	NA	NA	NA	215	259	349
8	NA	NA	NA	NA	197	226	264	352
10	NA	NA	NA	NA	214	245	278	358
15	NA	NA	NA	NA	NA	296	331	398
20	NA	NA	NA	NA	NA	352	387	457
30	NA	NA	NA	NA	NA	NA	507	581
50	NA	NA	NA	NA	NA	NA	NA	NA
5 to 16°F	Local 99% Winter Design Temperature: 5 to 16°F							
6	NA	NA	NA	NA	NA	NA	NA	416
8	NA	NA	NA	NA	NA	NA	312	423
10	NA	NA	NA	NA	NA	289	331	430
15	NA	NA	NA	NA	NA	NA	393	485
20	NA	NA	NA	NA	NA	NA	450	547
30	NA	NA	NA	NA	NA	NA	NA	682
50	NA	NA	NA	NA	NA	NA	NA	972
-10 to 4°F	Local 99% Winter Design Temperature: -10 to 4°F							
6	NA	NA	NA	NA	NA	NA	NA	484
8	NA	NA	NA	NA	NA	NA	NA	494
10	NA	NA	NA	NA	NA	NA	NA	513
15	NA	NA	NA	NA	NA	NA	NA	586
20	NA	NA	NA	NA	NA	NA	NA	650
30	NA	NA	NA	NA	NA	NA	NA	805
50	NA	NA	NA	NA	NA	NA	NA	1,003
-11°F or Lower	Local 99% Winter Design Temperature: -11°F or Lower							
	Not recommended for any vent configurations							

Note: See Figure B-19 in Appendix B for a map showing local 99 percent winter design temperatures in the United States.

For SI: °C = [(°F - 32]/1.8, 1 inch = 25.4 mm, 1 foot = 304.8 mm, 1 British thermal unit per hour = 0.2931 W.

504.2.8 Vent area and diameter. Where the vertical vent has a larger diameter than the vent connector, the vertical vent diameter shall be used to determine the minimum vent capacity, and the connector diameter shall be used to determine the maximum vent capacity. The flow area of the vertical vent shall not exceed seven times the flow area of the listed appliance categorized vent area, flue collar area, or draft hood outlet area unless designed in accordance with approved engineering methods.

504.2.9 Chimney and vent locations. Tables 504.2(1), 504.2(2), 504.2(3), 504.2(4) and 504.2(5) shall only be used for chimneys and vents not exposed to the outdoors below the roof line. A Type B vent or listed chimney lining system passing through an unused masonry chimney flue shall not be considered to be exposed to the outdoors. A Type B vent shall not be considered to be exposed to the outdoors where it passes through an unventilated enclosure or chase insulated to a value of not less than R8.

Table 504.2(3) in combination with Table 504.2(6) shall be used for clay-tile-lined exterior masonry chimneys, provided that all of the following are met:

1. Vent connector is a Type B double wall.

2. Vent connector length is limited to $1^1/_2$ feet for each inch (18 mm per mm) of vent connector diameter.

3. The appliance is draft hood equipped.

4. The input rating is less than the maximum capacity given by Table 504.2(3).

5. For a water heater, the outdoor design temperature is not less than 5°F (-15°C).

6. For a space-heating appliance, the input rating is greater than the minimum capacity given by Table 504.2(6).

Where these conditions cannot be met, an alternative venting design shall be used, such as a listed chimney lining system.

Exception: The installation of vents serving listed appliances shall be permitted to be in accordance with the appliance manufacturer's installation instructions.

504.2.10 Corrugated vent connector size. Corrugated vent connectors shall be not smaller than the listed appliance categorized vent diameter, flue collar diameter, or draft hood outlet diameter.

504.2.11 Vent connector size limitation. Vent connectors shall not be increased in size more than two sizes greater than the listed appliance categorized vent diameter, flue collar diameter, or draft hood outlet diameter.

504.2.12 Component commingling. In a single run of vent or vent connector, different diameters and types of vent and connector components shall be permitted to be used, provided that all such sizes and types are permitted by the tables.

504.2.13 Draft hood conversion accessories. Draft hood conversion accessories for use with masonry chimneys venting listed Category I fan-assisted appliances shall be listed and installed in accordance with the manufacturer's installation instructions for such listed accessories.

504.2.14 Table interpolation. Interpolation shall be permitted in calculating capacities for vent dimensions that fall between the table entries (see Example 3, Appendix B).

504.2.15 Extrapolation prohibited. Extrapolation beyond the table entries shall not be permitted.

504.2.16 Engineering calculations. For vent heights less than 6 feet (1829 mm) and greater than shown in the tables, engineering methods shall be used to calculate vent capacities.

504.3 Application of multiple appliance vent Tables 504.3(1) through 504.3(7). The application of Tables 504.3(1) through 504.3(7) shall be subject to the requirements of Sections 504.3.1 through 504.3.27.

504.3.1 Vent obstructions. These venting tables shall not be used where obstructions, as described in Section 503.15, are installed in the venting system. The installation of vents serving listed appliances with vent dampers shall be in accordance with the appliance manufacturer's instructions or in accordance with the following:

1. The maximum capacity of the vent connector shall be determined using the NAT Max column.

2. The maximum capacity of the vertical vent or chimney shall be determined using the FAN+NAT column when the second appliance is a fan-assisted appliance, or the NAT+NAT column when the second appliance is equipped with a draft hood.

3. The minimum capacity shall be determined as if the appliance were a fan-assisted appliance.

   3.1. The minimum capacity of the vent connector shall be determined using the FAN Min column.

   3.2. The FAN+FAN column shall be used where the second appliance is a fan-assisted appliance, and the FAN+NAT column shall be used where the second appliance is equipped with a draft hood, to determine whether the vertical vent or chimney configuration is not permitted (NA). Where the vent configuration is NA, the vent configuration shall not be permitted and an alternative venting configuration shall be utilized.

504.3.2 Connector length limit. The vent connector shall be routed to the vent utilizing the shortest possible route. Except as provided in Section 504.3.3, the maximum vent connector horizontal length shall be $1^1/_2$ feet for each inch (457 mm per mm) of connector diameter as shown in Table 504.3.2.

504.3.3 Connectors with longer lengths. Connectors with longer horizontal lengths than those listed in Section 504.3.2 are permitted under the following conditions:

1. The maximum capacity (FAN Max or NAT Max) of the vent connector shall be reduced 10 percent for each additional multiple of the length allowed by Sec-

tion 504.3.2. For example, the maximum length listed in Table 504.3.2 for a 4-inch (102 mm) connector is 6 feet (1829 mm). With a connector length greater than 6 feet (1829 mm) but not exceeding 12 feet (3658 mm), the maximum capacity must be reduced by 10 percent (0.90 × maximum vent connector capacity). With a connector length greater than 12 feet (3658 mm) but not exceeding 18 feet (5486 mm), the maximum capacity must be reduced by 20 percent (0.80 × maximum vent capacity).

2. For a connector serving a fan-assisted appliance, the minimum capacity (FAN Min) of the connector shall be determined by referring to the corresponding single appliance table. For Type B double-wall connectors, Table 504.2(1) shall be used. For single-wall connectors, Table 504.2(2) shall be used. The height (H) and lateral (L) shall be measured according to the procedures for a single-appliance vent, as if the other appliances were not present.

TABLE 504.3.2
MAXIMUM VENT CONNECTOR LENGTH

CONNECTOR DIAMETER MAXIMUM (inches)	CONNECTOR HORIZONTAL LENGTH (feet)
3	$4^1/_2$
4	6
5	$7^1/_2$
6	9
7	$10^1/_2$
8	12
9	$13^1/_2$
10	15
12	18
14	21
16	24
18	27
20	30
22	33
24	36

For SI: 1 inch = 25.4 mm, 1 foot = 304.8 mm.

504.3.4 Vent connector manifold. Where the vent connectors are combined prior to entering the vertical portion of the common vent to form a common vent manifold, the size of the common vent manifold and the common vent shall be determined by applying a 10-percent reduction (0.90 × maximum common vent capacity) to the common vent capacity part of the common vent tables. The length of the common vent connector manifold (L_m) shall not exceed $1^1/_2$ feet for each inch (457 mm per mm) of common vent connector manifold diameter (D) (see Figure B-11).

504.3.5 Common vertical vent offset. Where the common vertical vent is offset, the maximum capacity of the common vent shall be reduced in accordance with Section 504.3.6. The horizontal length of the common vent offset (L_o) shall

not exceed $1^1/_2$ feet for each inch (457 mm per mm) of common vent diameter.

504.3.6 Elbows in vents. For each elbow up to and including 45 degrees (0.79 rad) in the common vent, the maximum common vent capacity listed in the venting tables shall be reduced by 5 percent. For each elbow greater than 45 degrees (0.79 rad) up to and including 90 degrees (1.57 rad), the maximum common vent capacity listed in the venting tables shall be reduced by 10 percent.

504.3.7 Elbows in connectors. The vent connector capacities listed in the common vent sizing tables include allowance for two 90-degree (1.57 rad) elbows. For each additional elbow up to and including 45 degrees (0.79 rad), the maximum vent connector capacity listed in the venting tables shall be reduced by 5 percent. For each elbow greater than 45 degrees (0.79 rad) up to and including 90 degrees (1.57 rad), the maximum vent connector capacity listed in the venting tables shall be reduced by 10 percent.

504.3.8 Common vent minimum size. The cross-sectional area of the common vent shall be equal to or greater than the cross-sectional area of the largest connector.

504.3.9 Common vent fittings. At the point where tee or wye fittings connect to a common vent, the opening size of the fitting shall be equal to the size of the common vent. Such fittings shall not be prohibited from having reduced-size openings at the point of connection of appliance vent connectors.

504.3.9.1 Tee and wye fittings. Tee and wye fittings connected to a common vent shall be considered as part of the common vent and shall be constructed of materials consistent with that of the common vent.

504.3.10 High-altitude installations. Sea-level input ratings shall be used when determining maximum capacity for high-altitude installation. Actual input (derated for altitude) shall be used for determining minimum capacity for high-altitude installation.

504.3.11 Connector rise measurement. Connector rise (R) for each appliance connector shall be measured from the draft hood outlet or flue collar to the centerline where the vent gas streams come together.

504.3.12 Vent height measurement. For multiple appliances all located on one floor, available total height (H) shall be measured from the highest draft hood outlet or flue collar up to the level of the outlet of the common vent.

504.3.13 Multistory height measurement. For multistory installations, available total height (H) for each segment of the system shall be the vertical distance between the highest draft hood outlet or flue collar entering that segment and the centerline of the next higher interconnection tee (see Figure B-13).

504.3.14 Multistory lowest portion sizing. The size of the lowest connector and of the vertical vent leading to the lowest interconnection of a multistory system shall be in accordance with Table 504.2(1) or 504.2(2) for available total height (H) up to the lowest interconnection (see Figure B-14).

504.3.15 Multistory common vents. Where used in multistory systems, vertical common vents shall be Type B double wall and shall be installed with a listed vent cap.

504.3.16 Multistory common vent offsets. Offsets in multistory common vent systems shall be limited to a single offset in each system, and systems with an offset shall comply with all of the following:

1. The offset angle shall not exceed 45 degrees (0.79 rad) from vertical.

2. The horizontal length of the offset shall not exceed $1^{1}/_{2}$ feet for each inch (457 mm per mm) of common vent diameter of the segment in which the offset is located.

3. For the segment of the common vertical vent containing the offset, the common vent capacity listed in the common venting tables shall be reduced by 20 percent (0.80 × maximum common vent capacity).

4. A multistory common vent shall not be reduced in size above the offset.

504.3.17 Vertical vent maximum size. Where two or more appliances are connected to a vertical vent or chimney, the flow area of the largest section of vertical vent or chimney shall not exceed seven times the smallest listed appliance categorized vent areas, flue collar area, or draft hood outlet area unless designed in accordance with approved engineering methods.

504.3.18 Multiple input rate appliances. For appliances with more than one input rate, the minimum vent connector capacity (FAN Min) determined from the tables shall be less than the lowest appliance input rating, and the maximum vent connector capacity (FAN Max or NAT Max) determined from the tables shall be greater than the highest appliance input rating.

504.3.19 Liner system sizing and connections. Listed, corrugated metallic chimney liner systems in masonry chimneys shall be sized by using Table 504.3(1) or 504.3(2) for Type B vents, with the maximum capacity reduced by 20 percent (0.80 × maximum capacity) and the minimum capacity as shown in Table 504.3(1) or 504.3(2). Corrugated metallic liner systems installed with bends or offsets shall have their maximum capacity further reduced in accordance with Sections 504.3.5 and 504.3.6. The 20-percent reduction for corrugated metallic chimney liner systems includes an allowance for one long-radius 90-degree (1.57 rad) turn at the bottom of the liner. Where double-wall connectors are required, tee and wye fittings used to connect to the common vent chimney liner shall be listed double-wall fittings. Connections between chimney liners and listed double-wall fittings shall be made with listed adapter fittings designed for such purpose.

504.3.20 Chimney and vent location. Tables 504.3(1), 504.3(2), 504.3(3), 504.3(4) and 504.3(5) shall only be used for chimneys and vents not exposed to the outdoors below the roof line. A Type B vent or listed chimney lining system passing through an unused masonry chimney flue shall not be considered to be exposed to the outdoors. A Type B vent shall not be considered to be exposed to the outdoors where

it passes through an unventilated enclosure or chase insulated to a value of not less than R8.

Tables 504.3(6) and 504.3(7) shall be used for clay-tile-lined exterior masonry chimneys, provided all of the following conditions are met:

1. Vent connector is Type B double wall.

2. At least one appliance is draft hood equipped.

3. The combined appliance input rating is less than the maximum capacity given by Table 504.3(6a) for NAT+NAT or Table 504.3(7a) for FAN+NAT.

4. The input rating of each space-heating appliance is greater than the minimum input rating given by Table 504.3(6b) for NAT+NAT or Table 504.3(7b) for FAN+NAT.

5. The vent connector sizing is in accordance with Table 504.3(3).

Where these conditions cannot be met, an alternative venting design shall be used, such as a listed chimney lining system.

Exception: Vents serving listed appliances installed in accordance with the appliance manufacturer's installation instructions.

504.3.21 Connector maximum and minimum size. Vent connectors shall not be increased in size more than two sizes greater than the listed appliance categorized vent diameter, flue collar diameter, or draft hood outlet diameter. Vent connectors for draft hood-equipped appliances shall not be smaller than the draft hood outlet diameter. Where a vent connector size(s) determined from the tables for a fan-assisted appliance(s) is smaller than the flue collar diameter, the use of the smaller size(s) shall be permitted provided that the installation complies with all of the following conditions:

1. Vent connectors for fan-assisted appliance flue collars 12 inches (305 mm) in diameter or smaller are not reduced by more than one table size [e.g., 12 inches to 10 inches (305 mm to 254 mm) is a one-size reduction] and those larger than 12 inches (305 mm) in diameter are not reduced more than two table sizes [e.g., 24 inches to 20 inches (610 mm to 508 mm) is a two-size reduction].

2. The fan-assisted appliance(s) is common vented with a draft-hood-equipped appliances(s).

3. The vent connector has a smooth interior wall.

504.3.22 Component commingling. All combinations of pipe sizes, single-wall and double-wall metal pipe shall be allowed within any connector run(s) or within the common vent, provided that all of the appropriate tables permit all of the desired sizes and types of pipe, as if they were used for the entire length of the subject connector or vent. Where single-wall and Type B double-wall metal pipes are used for vent connectors within the same venting system, the common vent must be sized using Table 504.3(2) or 504.3(4), as appropriate.

504.3.23 Draft hood conversion accessories. Draft hood conversion accessories for use with masonry chimneys venting listed Category I fan-assisted appliances shall be listed and installed in accordance with the manufacturer's installation instructions for such listed accessories.

504.3.24 Multiple sizes permitted. Where a table permits more than one diameter of pipe to be used for a connector or vent, all the permitted sizes shall be permitted to be used.

504.3.25 Table interpolation. Interpolation shall be permitted in calculating capacities for vent dimensions that fall between table entries (see Appendix B, Example 3).

504.3.26 Extrapolation prohibited. Extrapolation beyond the table entries shall not be permitted.

504.3.27 Engineering calculations. For vent heights less than 6 feet (1829 mm) and greater than shown in the tables, engineering methods shall be used to calculate vent capacities.

SECTION 505 (IFGC)
DIRECT-VENT, INTEGRAL VENT,
MECHANICAL VENT AND
VENTILATION/EXHAUST HOOD VENTING

505.1 General. The installation of direct-vent and integral vent appliances shall be in accordance with Section 503. Mechanical venting systems and exhaust hood venting systems shall be designed and installed in accordance with Section 503.

505.1.1 Commercial cooking appliances vented by exhaust hoods. Where commercial cooking appliances are vented by means of the Type I or II kitchen exhaust hood system that serves such appliances, the exhaust system shall be fan powered and the appliances shall be interlocked with the exhaust hood system to prevent appliance operation when the exhaust hood system is not operating. Where a solenoid valve is installed in the gas piping as part of an interlock system, gas piping shall not be installed to bypass such valve. Dampers shall not be installed in the exhaust system.

> **Exception:** An interlock between the cooking appliance(s) and the exhaust hood system shall not be required where heat sensors or other approved methods automatically activate the exhaust hood system when cooking operations occur.

SECTION 506 (IFGC)
FACTORY-BUILT CHIMNEYS

506.1 Building heating appliances. Factory-built chimneys for building heating appliances producing flue gases having a temperature not greater than 1,000°F (538°C), measured at the entrance to the chimney, shall be listed and labeled in accordance with UL 103 and shall be installed and terminated in accordance with the manufacturer's installation instructions.

506.2 Support. Where factory-built chimneys are supported by structural members, such as joists and rafters, such members shall be designed to support the additional load.

506.3 Medium-heat appliances. Factory-built chimneys for medium-heat appliances producing flue gases having a temperature above 1,000°F (538°C), measured at the entrance to the chimney, shall be listed and labeled in accordance with UL 959 and shall be installed and terminated in accordance with the manufacturer's installation instructions.

Number of Appliances	Two or more
Appliance Type	Category I
Appliance Vent Connection	Type B double-wall connector

TABLE 504.3(1)
TYPE B DOUBLE-WALL VENT

VENT CONNECTOR CAPACITY

VENT HEIGHT (H) (feet)	CONNECTOR RISE (R) (feet)	3 FAN Min	3 FAN Max	3 NAT Max	4 FAN Min	4 FAN Max	4 NAT Max	5 FAN Min	5 FAN Max	5 NAT Max	6 FAN Min	6 FAN Max	6 NAT Max	7 FAN Min	7 FAN Max	7 NAT Max	8 FAN Min	8 FAN Max	8 NAT Max	9 FAN Min	9 FAN Max	9 NAT Max	10 FAN Min	10 FAN Max	10 NAT Max
6	1	22	37	26	35	66	46	46	106	72	58	164	104	77	225	142	92	296	185	109	376	237	128	466	289
	2	23	41	31	37	75	55	48	121	86	60	183	124	79	253	168	95	333	220	112	424	282	131	526	345
	3	24	44	35	38	81	62	49	132	96	62	199	139	82	275	189	97	363	248	114	463	317	134	575	386
8	1	22	40	27	35	72	48	49	114	76	64	176	109	84	243	148	100	320	194	118	408	248	138	507	303
	2	23	44	32	36	80	57	51	128	90	66	195	129	86	269	175	103	356	230	121	454	294	141	564	358
	3	24	47	36	37	87	64	53	139	101	67	210	145	88	290	198	105	384	258	123	492	330	143	612	402
10	1	22	43	28	34	78	50	49	123	78	65	189	113	89	257	154	106	341	200	125	436	257	146	542	314
	2	23	47	33	36	86	59	51	136	93	67	206	134	91	282	182	109	374	238	128	479	305	149	596	372
	3	24	50	37	37	92	67	52	146	104	69	220	150	94	303	205	111	402	268	131	515	342	152	642	417
15	1	21	50	30	33	89	53	47	142	83	64	220	120	88	298	163	110	389	214	134	493	273	162	609	333
	2	22	53	35	35	96	63	49	153	99	66	235	142	91	320	193	112	419	253	137	532	323	165	658	394
	3	24	55	40	36	102	71	51	163	111	68	248	160	93	339	218	115	445	286	140	565	365	167	700	444
20	1	21	54	31	33	99	56	46	157	87	62	246	125	86	334	171	107	436	224	131	552	285	158	681	347
	2	22	57	37	34	105	66	48	167	104	64	259	149	89	354	202	110	463	265	134	587	339	161	725	414
	3	23	60	42	35	110	74	50	176	116	66	271	168	91	371	228	113	486	300	137	618	383	164	764	466
30	1	20	62	33	31	113	59	45	181	93	60	288	134	83	391	182	103	512	238	125	649	305	151	802	372
	2	21	64	39	33	118	70	47	190	110	62	299	158	85	408	215	105	535	282	129	679	360	155	840	439
	3	22	66	44	34	123	79	48	198	124	64	309	178	88	423	242	108	555	317	132	706	405	158	874	494
50	1	19	71	36	30	133	64	43	216	101	57	349	145	78	477	197	97	627	257	120	797	330	144	984	403
	2	21	73	43	32	137	76	45	223	119	59	358	172	81	490	234	100	645	306	123	820	392	148	1,014	478
	3	22	75	48	33	141	86	46	229	134	61	366	194	83	502	263	103	661	343	126	842	441	151	1,043	538
100	1	18	82	37	28	158	66	40	262	104	53	442	150	73	611	204	91	810	266	112	1,038	341	135	1,285	417
	2	19	83	44	30	161	79	42	267	123	55	447	178	75	619	242	94	822	316	115	1,054	405	139	1,306	494
	3	20	84	50	31	163	89	44	272	138	57	452	109	78	627	272	97	834	355	118	1,069	455	142	1,327	555

COMMON VENT CAPACITY

VENT HEIGHT (H) (feet)	4 FAN+FAN	4 FAN+NAT	4 NAT+NAT	5 FAN+FAN	5 FAN+NAT	5 NAT+NAT	6 FAN+FAN	6 FAN+NAT	6 NAT+NAT	7 FAN+FAN	7 FAN+NAT	7 NAT+NAT	8 FAN+FAN	8 FAN+NAT	8 NAT+NAT	9 FAN+FAN	9 FAN+NAT	9 NAT+NAT	10 FAN+FAN	10 FAN+NAT	10 NAT+NAT
6	92	81	65	140	116	103	204	161	147	309	248	200	404	314	260	547	434	335	672	520	410
8	101	90	73	155	129	114	224	178	163	339	275	223	444	348	290	602	480	378	740	577	465
10	110	97	79	169	141	124	243	194	178	367	299	242	477	377	315	649	522	405	800	627	495
15	125	112	91	195	164	144	283	228	206	427	352	280	556	444	365	753	612	465	924	733	565
20	136	123	102	215	183	160	314	255	229	475	394	310	621	499	405	842	688	523	1,035	826	640
30	152	138	118	244	210	185	361	297	266	547	459	360	720	585	470	979	808	605	1,209	975	740
50	167	153	134	279	244	214	421	353	310	641	547	423	854	706	550	1,164	977	705	1,451	1,188	860
100	175	163	NA	311	277	NA	489	421	NA	751	658	479	1,025	873	625	1,408	1,215	800	1,784	1,502	975

(continued)

Number of Appliances	Two or more
Appliance Type	Category I
Appliance Vent Connection	Type B double-wall connector

TABLE 504.3(1)—continued
TYPE B DOUBLE-WALL VENT

VENT CONNECTOR CAPACITY

VENT HEIGHT (H) (feet)	CONNECTOR RISE (R) (feet)	12 FAN Min	12 FAN Max	12 NAT Max	14 FAN Min	14 FAN Max	14 NAT Max	16 FAN Min	16 FAN Max	16 NAT Max	18 FAN Min	18 FAN Max	18 NAT Max	20 FAN Min	20 FAN Max	20 NAT Max	22 FAN Min	22 FAN Max	22 NAT Max	24 FAN Min	24 FAN Max	24 NAT Max
6	2	174	764	496	223	1,046	653	281	1,371	853	346	1,772	1,080	NA	NA	NA	NA	NA	NA	NA	NA	NA
6	4	180	897	616	230	1,231	827	287	1,617	1,081	352	2,069	1,370	NA	NA	NA	NA	NA	NA	NA	NA	NA
6	6	NA	NA	NA	NA	NA	NA	NA	NA	NA	NA	NA	NA	NA	NA	NA	NA	NA	NA	NA	NA	NA
8	2	186	822	516	238	1,126	696	298	1,478	910	365	1,920	1,150	NA	NA	NA	NA	NA	NA	NA	NA	NA
8	4	192	952	644	244	1,307	884	305	1,719	1,150	372	2,211	1,460	471	2,737	1,800	560	3,319	2,180	662	3,957	2,590
8	6	198	1,050	772	252	1,445	1,072	313	1,902	1,390	380	2,434	1,770	478	3,018	2,180	568	3,665	2,640	669	4,373	3,130
10	2	196	870	536	249	1,195	730	311	1,570	955	379	2,049	1,205	NA	NA	NA	NA	NA	NA	NA	NA	NA
10	4	201	997	664	256	1,371	924	318	1,804	1,205	387	2,332	1,535	486	2,887	1,890	581	3,502	2,280	686	4,175	2,710
10	6	207	1,095	792	263	1,509	1,118	325	1,989	1,455	395	2,556	1,865	494	3,169	2,290	589	3,849	2,760	694	4,593	3,270
15	2	214	967	568	272	1,334	790	336	1,760	1,030	408	2,317	1,305	NA	NA	NA	NA	NA	NA	NA	NA	NA
15	4	221	1,085	712	279	1,499	1,006	344	1,978	1,320	416	2,579	1,665	523	3,197	2,060	624	3,881	2,490	734	4,631	2,960
15	6	228	1,181	856	286	1,632	1,222	351	2,157	1,610	424	2,796	2,025	533	3,470	2,510	634	4,216	3,030	743	5,035	3,600
20	2	223	1,051	596	291	1,443	840	357	1,911	1,095	430	2,533	1,385	NA	NA	NA	NA	NA	NA	NA	NA	NA
20	4	230	1,162	748	298	1,597	1,064	365	2,116	1,395	438	2,778	1,765	554	3,447	2,180	661	4,190	2,630	772	5,005	3,130
20	6	237	1,253	900	307	1,726	1,288	373	2,287	1,695	450	2,984	2,145	567	3,708	2,650	671	4,511	3,190	785	5,392	3,790
30	2	216	1,217	632	286	1,664	910	367	2,183	1,190	461	2,891	1,540	NA	NA	NA	NA	NA	NA	NA	NA	NA
30	4	223	1,316	792	294	1,802	1,160	376	2,366	1,510	474	3,110	1,920	619	3,840	2,365	728	4,861	2,860	847	5,606	3,410
30	6	231	1,400	952	303	1,920	1,410	384	2,524	1,830	485	3,299	2,340	632	4,080	2,875	741	4,976	3,480	860	5,961	4,150
50	2	206	1,479	689	273	2,023	1,007	350	2,659	1,315	435	3,548	1,665	NA	NA	NA	NA	NA	NA	NA	NA	NA
50	4	213	1,561	860	281	2,139	1,291	359	2,814	1,685	447	3,730	2,135	580	4,601	2,633	709	5,569	3,185	851	6,633	3,790
50	6	221	1,631	1,031	290	2,242	1,575	369	2,951	2,055	461	3,893	2,605	594	4,808	3,208	724	5,826	3,885	867	6,943	4,620
100	2	192	1,923	712	254	2,644	1,050	326	3,490	1,370	402	4,707	1,740	NA	NA	NA	NA	NA	NA	NA	NA	NA
100	4	200	1,984	888	263	2,731	1,346	336	3,606	1,760	414	4,842	2,220	523	5,982	2,750	639	7,254	3,330	769	8,650	3,950
100	6	208	2,035	1,064	272	2,811	1,642	346	3,714	2,150	426	4,968	2,700	539	6,143	3,350	654	7,453	4,070	786	8,892	4,810

COMMON VENT CAPACITY

VENT HEIGHT (H) (feet)	12 FAN+FAN	12 FAN+NAT	12 NAT+NAT	14 FAN+FAN	14 FAN+NAT	14 NAT+NAT	16 FAN+FAN	16 FAN+NAT	16 NAT+NAT	18 FAN+FAN	18 FAN+NAT	18 NAT+NAT	20 FAN+FAN	20 FAN+NAT	20 NAT+NAT	22 FAN+FAN	22 FAN+NAT	22 NAT+NAT	24 FAN+FAN	24 FAN+NAT	24 NAT+NAT
6	900	696	588	1,284	990	815	1,735	1,336	1,065	2,253	1,732	1,345	2,838	2,180	1,660	3,488	2,677	1970	4,206	3,226	2,390
8	994	773	652	1,423	1,103	912	1,927	1,491	1,190	2,507	1,936	1,510	3,162	2,439	1,860	3,890	2,998	2,200	4,695	3,616	2,680
10	1,076	841	712	1,542	1,200	995	2,093	1,625	1,300	2,727	2,113	1645	3,444	2,665	2,030	4,241	3,278	2,400	5,123	3,957	2,920
15	1,247	986	825	1,794	1,410	1,158	2,440	1,910	1,510	3,184	2,484	1,910	4,026	3,133	2,360	4,971	3,862	2,790	6,016	4,670	3,400
20	1,405	1,116	916	2,006	1,588	1,290	2,722	2,147	1,690	3,561	2,798	2,140	4,548	3,552	2,640	5,573	4,352	3,120	6,749	5,261	3,800
30	1,658	1,327	1,025	2,373	1,892	1,525	3,220	2,558	1,990	4,197	3,326	2,520	5,303	4,193	3,110	6,539	5,157	3,680	7,940	6,247	4,480
50	2,024	1,640	1,280	2,911	2,347	1,863	3,964	3,183	2,430	5,184	4,149	3,075	6,567	5,240	3,800	8,116	6,458	4,500	9,837	7,813	5,475
100	2,569	2,131	1,670	3,732	3,076	2,450	5,125	4,202	3,200	6,749	5,509	4,050	8,597	6,986	5,000	10,681	8,648	5,920	13,004	10,499	7,200

For SI: 1 inch = 25.4 mm, 1 foot = 304.8 mm, 1 British thermal unit per hour = 0.2931 W.

TABLE 504.3(2)
TYPE B DOUBLE-WALL VENT

Number of Appliances	Two or more
Appliance Type	Category I
Appliance Vent Connection	Single-wall metal connector

VENT CONNECTOR CAPACITY

		SINGLE-WALL METAL VENT CONNECTOR DIAMETER—(*D*) inches																						
		3			4			5			6			7			8			9			10		
VENT HEIGHT (*H*) (feet)	CONNECTOR RISE (*R*) (feet)	APPLIANCE INPUT RATING LIMITS IN THOUSANDS OF BTU/H																							
		FAN		NAT	FAN		NAT	FAN		NAT	FAN		NAT	FAN		NAT	FAN		NAT	FAN		NAT	FAN		NAT
		Min	Max	Max	Min	Max	Max	Min	Max	Max	Min	Max	Max	Min	Max	Max	Min	Max	Max	Min	Max	Max	Min	Max	Max
6	1	NA	NA	26	NA	NA	46	NA	NA	71	NA	NA	102	207	223	140	262	293	183	325	373	234	447	463	286
	2	NA	NA	31	NA	NA	55	NA	NA	85	168	182	123	215	251	167	271	331	219	334	422	281	458	524	344
	3	NA	NA	34	NA	NA	62	121	131	95	175	198	138	222	273	188	279	361	247	344	462	316	468	574	385
8	1	NA	NA	27	NA	NA	48	NA	NA	75	NA	NA	106	226	240	145	285	316	191	352	403	244	481	502	299
	2	NA	NA	32	NA	NA	57	125	126	89	184	193	127	234	266	173	293	353	228	360	450	292	492	560	355
	3	NA	NA	35	NA	NA	64	130	138	100	191	208	144	241	287	197	302	381	256	370	489	328	501	609	400
10	1	NA	NA	28	NA	NA	50	119	121	77	182	186	110	240	253	150	302	335	196	372	429	252	506	534	308
	2	NA	NA	33	84	85	59	124	134	91	189	203	132	248	278	183	311	369	235	381	473	302	517	589	368
	3	NA	NA	36	89	91	67	129	144	102	197	217	148	257	299	203	320	398	265	391	511	339	528	637	413
15	1	NA	NA	29	79	87	52	116	138	81	177	214	116	238	291	158	312	380	208	397	482	266	556	596	324
	2	NA	NA	34	83	94	62	121	150	97	185	230	138	246	314	189	321	411	248	407	522	317	568	646	387
	3	NA	NA	39	87	100	70	127	160	109	193	243	157	255	333	215	331	438	281	418	557	360	579	690	437
20	1	49	56	30	78	97	54	115	152	84	175	238	120	233	325	165	306	425	217	390	538	276	546	664	336
	2	52	59	36	82	103	64	120	163	101	182	252	144	243	346	197	317	453	259	400	574	331	558	709	403
	3	55	62	40	87	107	72	125	172	113	190	264	164	252	363	223	326	476	294	412	607	375	570	750	457
30	1	47	60	31	77	110	57	112	175	89	169	278	129	226	380	175	296	497	230	378	630	294	528	779	358
	2	51	62	37	81	115	67	117	185	106	177	290	152	236	397	208	307	521	274	389	662	349	541	819	425
	3	54	64	42	85	119	76	122	193	120	185	300	172	244	412	235	316	542	309	400	690	394	555	855	482
50	1	46	69	34	75	128	60	109	207	96	162	336	137	217	460	188	284	604	245	364	768	314	507	951	384
	2	49	71	40	79	132	72	114	215	113	170	345	164	226	473	223	294	623	293	376	793	375	520	983	458
	3	52	72	45	83	136	82	119	221	123	178	353	186	235	486	252	304	640	331	387	816	423	535	1,013	518
100	1	45	79	34	71	150	61	104	249	98	153	424	140	205	585	192	269	774	249	345	993	321	476	1,236	393
	2	48	80	41	75	153	73	110	255	115	160	428	167	212	593	228	279	788	299	358	1,011	383	490	1,259	469
	3	51	81	46	79	157	85	114	260	129	168	433	190	222	603	256	289	801	339	368	1,027	431	506	1,280	527

COMMON VENT CAPACITY

	TYPE B DOUBLE-WALL COMMON VENT DIAMETER— (*D*) inches																				
	4			5			6			7			8			9			10		
VENT HEIGHT (*H*) (feet)	COMBINED APPLIANCE INPUT RATING IN THOUSANDS OF BTU/H																				
	FAN +FAN	FAN +NAT	NAT +NAT	FAN +FAN	FAN +NAT	NAT +NAT	FAN +FAN	FAN +NAT	NAT +NAT	FAN +FAN	FAN +NAT	NAT +NAT	FAN +FAN	FAN +NAT	NAT +NAT	FAN +FAN	FAN +NAT	NAT +NAT	FAN +FAN	FAN +NAT	NAT +NAT
6	NA	78	64	NA	113	99	200	158	144	304	244	196	398	310	257	541	429	332	665	515	407
8	NA	87	71	NA	126	111	218	173	159	331	269	218	436	342	285	592	473	373	730	569	460
10	NA	94	76	163	137	120	237	189	174	357	292	236	467	369	309	638	512	398	787	617	487
15	121	108	88	189	159	140	275	221	200	416	343	274	544	434	357	738	599	456	905	718	553
20	131	118	98	208	177	156	305	247	223	463	383	302	606	487	395	824	673	512	1,013	808	626
30	145	132	113	236	202	180	350	286	257	533	446	349	703	570	459	958	790	593	1,183	952	723
50	159	145	128	268	233	208	406	337	296	622	529	410	833	686	535	1,139	954	689	1,418	1,157	838
100	166	153	NA	297	263	NA	469	398	NA	726	633	464	999	846	606	1,378	1,185	780	1,741	1,459	948

For SI: 1 inch = 25.4 mm, 1 foot = 304.8 mm, 1 British thermal unit per hour = 0.2931 W.

2006 INTERNATIONAL FUEL GAS CODE®

TABLE 504.3(3)
MASONRY CHIMNEY

Number of Appliances	Two or more
Appliance Type	Category I
Appliance Vent Connection	Type B double-wall connector

VENT CONNECTOR CAPACITY

VENT HEIGHT (H) (feet)	CONNECTOR RISE (R) (feet)	TYPE B DOUBLE-WALL VENT CONNECTOR DIAMETER—(D) inches																						
		3			4			5			6			7			8			9			10		
		APPLIANCE INPUT RATING LIMITS IN THOUSANDS OF BTU/H																						
		FAN		NAT	FAN		NAT	FAN		NAT	FAN		NAT	FAN		NAT	FAN		NAT	FAN		NAT	FAN	NAT
		Min	Max	Max	Min	Max	Max	Min	Max	Max	Min	Max	Max	Min	Max	Max	Min	Max	Max	Min	Max	Max	Min	Max	Max
6	1	24	33	21	39	62	40	52	106	67	65	194	101	87	274	141	104	370	201	124	479	253	145	599	319
	2	26	43	28	41	79	52	53	133	85	67	230	124	89	324	173	107	436	232	127	562	300	148	694	378
	3	27	49	34	42	92	61	55	155	97	69	262	143	91	369	203	109	491	270	129	633	349	151	795	439
8	1	24	39	22	39	72	41	55	117	69	71	213	105	94	304	148	113	414	210	134	539	267	156	682	335
	2	26	47	29	40	87	53	57	140	86	73	246	127	97	350	179	116	473	240	137	615	311	160	776	394
	3	27	52	34	42	97	62	59	159	98	75	269	145	99	383	206	119	517	276	139	672	358	163	848	452
10	1	24	42	22	38	80	42	55	130	71	74	232	108	101	324	153	120	444	216	142	582	277	165	739	348
	2	26	50	29	40	93	54	57	153	87	76	261	129	103	366	184	123	498	247	145	652	321	168	825	407
	3	27	55	35	41	105	63	58	170	100	78	284	148	106	397	209	126	540	281	147	705	366	171	893	463
15	1	24	48	23	38	93	44	54	154	74	72	277	114	100	384	164	125	511	229	153	658	297	184	824	375
	2	25	55	31	39	105	55	56	174	89	74	299	134	103	419	192	128	558	260	156	718	339	187	900	432
	3	26	59	35	41	115	64	57	189	102	76	319	153	105	448	215	131	597	292	159	760	382	190	960	486
20	1	24	52	24	37	102	46	53	172	77	71	313	119	98	437	173	123	584	239	150	752	312	180	943	397
	2	25	58	31	39	114	56	55	190	91	73	335	138	101	467	199	126	625	270	153	805	354	184	1,011	452
	3	26	63	35	40	123	65	57	204	104	75	353	157	104	493	222	129	661	301	156	851	396	187	1,067	505
30	1	24	54	25	37	111	48	52	192	82	69	357	127	96	504	187	119	680	255	145	883	337	175	1,115	432
	2	25	60	32	38	122	58	54	208	95	72	376	145	99	531	209	122	715	287	149	928	378	179	1,171	484
	3	26	64	36	40	131	66	56	221	107	74	392	163	101	554	233	125	746	317	152	968	418	182	1,220	535
50	1	23	51	25	36	116	51	51	209	89	67	405	143	92	582	213	115	798	294	140	1,049	392	168	1,334	506
	2	24	59	32	37	127	61	53	225	102	70	421	161	95	604	235	118	827	326	143	1,085	433	172	1,379	558
	3	26	64	36	39	135	69	55	237	115	72	435	80	98	624	260	121	854	357	147	1,118	474	176	1,421	611
100	1	23	46	24	35	108	50	49	208	92	65	428	155	88	640	237	109	907	334	134	1,222	454	161	1,589	596
	2	24	53	31	37	120	60	51	224	105	67	444	174	92	660	260	113	933	368	138	1,253	497	165	1,626	651
	3	25	59	35	38	130	68	53	237	118	69	458	193	94	679	285	116	956	399	141	1,282	540	169	1,661	705

COMMON VENT CAPACITY

VENT HEIGHT (H) (feet)	MINIMUM INTERNAL AREA OF MASONRY CHIMNEY FLUE (square inches)																							
	12			19			28			38			50			63			78			113		
	COMBINED APPLIANCE INPUT RATING IN THOUSANDS OF BTU/H																							
	FAN +FAN	FAN +NAT	NAT +NAT	FAN +FAN	FAN +NAT	NAT +NAT	FAN +FAN	FAN +NAT	NAT +NAT	FAN +FAN	FAN +NAT	NAT +NAT	FAN +FAN	FAN +NAT	NAT +NAT	FAN +FAN	FAN +NAT	NAT +NAT	FAN +FAN	FAN +NAT	NAT +NAT	FAN +FAN	FAN +NAT	NAT +NAT
6	NA	74	25	NA	119	46	NA	178	71	NA	257	103	NA	351	143	NA	458	188	NA	582	246	1,041	853	NA
8	NA	80	28	NA	130	53	NA	193	82	NA	279	119	NA	384	163	NA	501	218	724	636	278	1,144	937	408
10	NA	84	31	NA	138	56	NA	207	90	NA	299	131	NA	409	177	606	538	236	776	686	302	1,226	1,010	454
15	NA	NA	36	NA	152	67	NA	233	106	NA	334	152	523	467	212	682	611	283	874	781	365	1,374	1,156	546
20	NA	NA	41	NA	NA	75	NA	250	122	NA	368	172	565	508	243	742	668	325	955	858	419	1,513	1,286	648
30	NA	NA	NA	NA	NA	NA	NA	270	137	NA	404	198	615	564	278	816	747	381	1,062	969	496	1,702	1,473	749
50	NA	NA	NA	NA	NA	NA	NA	NA	NA	NA	NA	NA	NA	620	328	879	831	461	1,165	1,089	606	1,905	1,692	922
100	NA	NA	NA	NA	NA	NA	NA	NA	NA	NA	NA	NA	NA	NA	348	NA	NA	499	NA	NA	669	2,053	1,921	1,058

For SI: 1 inch = 25.4 mm, 1 square inch = 645.16 mm², 1 foot = 304.8 mm, 1 British thermal unit per hour = 0.2931 W.

Number of Appliances	Two or more
Appliance Type	Category I
Appliance Vent Connection	Single-wall metal connector

TABLE 504.3(4) MASONRY CHIMNEY

VENT CONNECTOR CAPACITY

VENT HEIGHT (*H*) (feet)	CONNECTOR RISE (*R*) (feet)	SINGLE-WALL METAL VENT CONNECTOR DIAMETER (*D*)—inches																							
		3			4			5			6			7			8			9			10		
		APPLIANCE INPUT RATING LIMITS IN THOUSANDS OF BTU/H																							
		FAN		NAT	FAN		NAT	FAN		NAT	FAN		NAT	FAN		NAT	FAN		NAT	FAN		NAT	FAN		NAT
		Min	Max	Max	Min	Max	Max	Min	Max	Max	Min	Max	Max	Min	Max	Max	Min	Max	Max	Min	Max	Max	Min	Max	Max
6	1	NA	NA	21	NA	NA	39	NA	NA	66	179	191	100	231	271	140	292	366	200	362	474	252	499	594	316
	2	NA	NA	28	NA	NA	52	NA	NA	84	186	227	123	239	321	172	301	432	231	373	557	299	509	696	376
	3	NA	NA	34	NA	NA	61	134	153	97	193	258	142	247	365	202	309	491	269	381	634	348	519	793	437
8	1	NA	NA	21	NA	NA	40	NA	NA	68	195	208	103	250	298	146	313	407	207	387	530	263	529	672	331
	2	NA	NA	28	NA	NA	52	137	139	85	202	240	125	258	343	177	323	465	238	397	607	309	540	766	391
	3	NA	NA	34	NA	NA	62	143	156	98	210	264	145	266	376	205	332	509	274	407	663	356	551	838	450
10	1	NA	NA	22	NA	NA	41	130	151	70	202	225	106	267	316	151	333	434	213	410	571	273	558	727	343
	2	NA	NA	29	NA	NA	53	136	150	86	210	255	128	276	358	181	343	489	244	420	640	317	569	813	403
	3	NA	NA	34	97	102	62	143	166	99	217	277	147	284	389	207	352	530	279	430	694	363	580	880	459
15	1	NA	NA	23	NA	NA	43	129	151	73	199	271	112	268	376	161	349	502	225	445	646	291	623	808	366
	2	NA	NA	30	92	103	54	135	170	88	207	295	132	277	411	189	359	548	256	456	706	334	634	884	424
	3	NA	NA	34	96	112	63	141	185	101	215	315	151	286	439	213	368	586	289	466	755	378	646	945	479
20	1	NA	NA	23	87	99	45	128	167	76	197	303	117	265	425	169	345	569	235	439	734	306	614	921	347
	2	NA	NA	30	91	111	55	134	185	90	205	325	136	274	455	195	355	610	266	450	787	348	627	986	443
	3	NA	NA	35	96	119	64	140	199	103	213	343	154	282	481	219	365	644	298	461	831	391	639	1,042	496
30	1	NA	NA	24	86	108	47	126	187	80	193	347	124	259	492	183	338	665	250	430	864	330	600	1,089	421
	2	NA	NA	31	91	119	57	132	203	93	201	366	142	269	518	205	348	699	282	442	908	372	613	1,145	473
	3	NA	NA	35	95	127	65	138	216	105	209	381	160	277	540	229	358	729	312	452	946	412	626	1,193	524
50	1	NA	NA	24	85	113	50	124	204	87	188	392	139	252	567	208	328	778	287	417	1,022	383	582	1,302	492
	2	NA	NA	31	89	123	60	130	218	100	196	408	158	262	588	230	339	806	320	429	1,058	425	596	1,346	545
	3	NA	NA	35	94	131	68	136	231	112	205	422	176	271	607	255	349	831	351	440	1,090	466	610	1,386	597
100	1	NA	NA	23	84	104	49	122	200	89	182	410	151	243	617	232	315	875	328	402	1,181	444	560	1,537	580
	2	NA	NA	30	88	115	59	127	215	102	190	425	169	253	636	254	326	899	361	415	1,210	488	575	1,570	634
	3	NA	NA	34	93	124	67	133	228	115	199	438	188	262	654	279	337	921	392	427	1,238	529	589	1,604	687

COMMON VENT CAPACITY

VENT HEIGHT (*H*) (feet)	MINIMUM INTERNAL AREA OF MASONRY CHIMNEY FLUE (square inches)																							
	12			19			28			38			50			63			78			113		
	COMBINED APPLIANCE INPUT RATING IN THOUSANDS OF BTU/H																							
	FAN +FAN	FAN +NAT	NAT +NAT	FAN +FAN	FAN +NAT	NAT +NAT	FAN +FAN	FAN +NAT	NAT +NAT	FAN +FAN	FAN +NAT	NAT +NAT	FAN +FAN	FAN +NAT	NAT +NAT	FAN +FAN	FAN +NAT	NAT +NAT	FAN +FAN	FAN +NAT	NAT +NAT	FAN +FAN	FAN +NAT	NAT +NAT
6	NA	NA	25	NA	118	45	NA	176	71	NA	255	102	NA	348	142	NA	455	187	NA	579	245	NA	846	NA
8	NA	NA	28	NA	128	52	NA	190	81	NA	276	118	NA	380	162	NA	497	217	NA	633	277	1,136	928	405
10	NA	NA	31	NA	136	56	NA	205	89	NA	295	129	NA	405	175	NA	532	234	171	680	300	1,216	1,000	450
15	NA	NA	36	NA	NA	66	NA	230	105	NA	335	150	NA	400	210	677	602	280	866	772	360	1,359	1,139	540
20	NA	NA	NA	NA	NA	74	NA	247	120	NA	362	170	NA	503	240	765	661	321	947	849	415	1,495	1,264	640
30	NA	NA	NA	NA	NA	NA	NA	NA	135	NA	398	195	NA	558	275	808	739	377	1,052	957	490	1,682	1,447	740
50	NA	NA	NA	NA	NA	NA	NA	NA	NA	NA	NA	NA	NA	612	325	NA	821	456	1,152	1,076	600	1,879	1,672	910
100	NA	NA	NA	NA	NA	NA	NA	NA	NA	NA	NA	NA	NA	NA	NA	NA	NA	494	NA	NA	663	2,006	1,885	1,046

For SI: 1 inch = 25.4 mm, 1 square inch = 645.16 mm^2, 1 foot = 304.8 mm, 1 British thermal unit per hour = 0.2931 W.

TABLE 504.3(5)
SINGLE-WALL METAL PIPE OR TYPE ASBESTOS CEMENT VENT

Number of Appliances	Two or more
Appliance Type	Draft hood-equipped
Appliance Vent Connection	Direct to pipe or vent

VENT CONNECTOR CAPACITY

TOTAL VENT HEIGHT (*H*) (feet)	CONNECTOR RISE (*R*) (feet)	VENT CONNECTOR DIAMETER—(*D*) inches					
		3	4	5	6	7	8
		MAXIMUM APPLIANCE INPUT RATING IN THOUSANDS OF BTU/H					
6-8	1	21	40	68	102	146	205
	2	28	53	86	124	178	235
	3	34	61	98	147	204	275
15	1	23	44	77	117	179	240
	2	30	56	92	134	194	265
	3	35	64	102	155	216	298
30 and up	1	25	49	84	129	190	270
	2	31	58	97	145	211	295
	3	36	68	107	164	232	321

COMMON VENT CAPACITY

TOTAL VENT HEIGHT (*H*) (feet)	COMMON VENT DIAMETER—(*D*) inches						
	4	5	6	7	8	10	12
	COMBINED APPLIANCE INPUT RATING IN THOUSANDS OF BTU/H						
6	48	78	111	155	205	320	NA
8	55	89	128	175	234	365	505
10	59	95	136	190	250	395	560
15	71	115	168	228	305	480	690
20	80	129	186	260	340	550	790
30	NA	147	215	300	400	650	940
50	NA	NA	NA	360	490	810	1,190

For SI: 1 inch = 25.4 mm, 1 foot = 304.8 mm, 1 British thermal unit per hour = 0.2931 W.

TABLE 504.3(6a)
EXTERIOR MASONRY CHIMNEY

Number of Appliances	Two or more
Appliance Type	NAT + NAT
Appliance Vent Connection	Type B double-wall connector

Combined Appliance Maximum Input Rating in Thousands of Btu per Hour

VENT HEIGHT (feet)	INTERNAL AREA OF CHIMNEY (square inches)							
	12	19	28	38	50	63	78	113
6	25	46	71	103	143	188	246	NA
8	28	53	82	119	163	218	278	408
10	31	56	90	131	177	236	302	454
15	NA	67	106	152	212	283	365	546
20	NA	NA	NA	NA	NA	325	419	648
30	NA	NA	NA	NA	NA	NA	496	749
50	NA	NA	NA	NA	NA	NA	NA	922
100	NA	NA	NA	NA	NA	NA	NA	NA

TABLE 504.3(6b)
EXTERIOR MASONRY CHIMNEY

Number of Appliances	Two or more
Appliance Type	NAT + NAT
Appliance Vent Connection	Type B double-wall connector

Minimum Allowable Input Rating of Space-Heating Appliance in Thousands of Btu per Hour

VENT HEIGHT (feet)	INTERNAL AREA OF CHIMNEY (square inches)							
	12	19	28	38	50	63	78	113
37°F or Greater	Local 99% Winter Design Temperature: 37°F or Greater							
6	0	0	0	0	0	0	0	NA
8	0	0	0	0	0	0	0	0
10	0	0	0	0	0	0	0	0
15	NA	0	0	0	0	0	0	0
20	NA	NA	NA	NA	NA	184	0	0
30	NA	NA	NA	NA	NA	393	334	0
50	NA	NA	NA	NA	NA	NA	NA	579
100	NA	NA	NA	NA	NA	NA	NA	NA
27 to 36°F	Local 99% Winter Design Temperature: 27 to 36°F							
6	0	0	68	NA	NA	180	212	NA
8	0	0	82	NA	NA	187	214	263
10	0	51	NA	NA	NA	201	225	265
15	NA	NA	NA	NA	NA	253	274	305
20	NA	NA	NA	NA	NA	307	330	362
30	NA	NA	NA	NA	NA	NA	445	485
50	NA	NA	NA	NA	NA	NA	NA	763
100	NA	NA	NA	NA	NA	NA	NA	NA

TABLE 504.3(6b)
EXTERIOR MASONRY CHIMNEY—continued

Minimum Allowable Input Rating of Space-Heating Appliance in Thousands of Btu per Hour

VENT HEIGHT (feet)	INTERNAL AREA OF CHIMNEY (square inches)							
	12	19	28	38	50	63	78	113
17 to 26°F	Local 99% Winter Design Temperature: 17 to 26°F							
6	NA	NA	NA	NA	NA	NA	NA	NA
8	NA	NA	NA	NA	NA	NA	264	352
10	NA	NA	NA	NA	NA	NA	278	358
15	NA	NA	NA	NA	NA	NA	331	398
20	NA	NA	NA	NA	NA	NA	387	457
30	NA	NA	NA	NA	NA	NA	NA	581
50	NA	NA	NA	NA	NA	NA	NA	862
100	NA	NA	NA	NA	NA	NA	NA	NA
5 to 16°F	Local 99% Winter Design Temperature: 5 to 16°F							
6	NA	NA	NA	NA	NA	NA	NA	NA
8	NA	NA	NA	NA	NA	NA	NA	NA
10	NA	NA	NA	NA	NA	NA	NA	430
15	NA	NA	NA	NA	NA	NA	NA	485
20	NA	NA	NA	NA	NA	NA	NA	547
30	NA	NA	NA	NA	NA	NA	NA	682
50	NA	NA	NA	NA	NA	NA	NA	NA
100	NA	NA	NA	NA	NA	NA	NA	NA
4°F or Lower	Local 99% Winter Design Temperature: 4°F or Lower							
	Not recommended for any vent configurations							

Note: See Figure B-19 in Appendix B for a map showing local 99 percent winter design temperatures in the United States.

For SI: °C = [(°F - 32)/1.8, 1 inch = 25.4 mm, 1 square inch = 645.16 mm², 1 foot = 304.8 mm, 1 British thermal unit per hour = 0.2931 W.

TABLE 504.3(7a)
EXTERIOR MASONRY CHIMNEY

Number of Appliances	Two or more
Appliance Type	FAN + NAT
Appliance Vent Connection	Type B double-wall connector

Combined Appliance Maximum Input Rating in Thousands of Btu per Hour

VENT HEIGHT (feet)	INTERNAL AREA OF CHIMNEY (square inches)							
	12	19	28	38	50	63	78	113
6	74	119	178	257	351	458	582	853
8	80	130	193	279	384	501	636	937
10	84	138	207	299	409	538	686	1,010
15	NA	152	233	334	467	611	781	1,156
20	NA	NA	250	368	508	668	858	1,286
30	NA	NA	NA	404	564	747	969	1,473
50	NA	NA	NA	NA	NA	831	1,089	1,692
100	NA	NA	NA	NA	NA	NA	NA	1,921

TABLE 504.3(7b)
EXTERIOR MASONRY CHIMNEY

Number of Appliances	Two or more
Appliance Type	FAN + NAT
Appliance Vent Connection	Type B double-wall connector

Minimum Allowable Input Rating of Space-Heating Appliance in Thousands of Btu per Hour

VENT HEIGHT (feet)	INTERNAL AREA OF CHIMNEY (square inches)							
	12	19	28	38	50	63	78	113
37°F or Greater	Local 99% Winter Design Temperature: 37°F or Greater							
6	0	0	0	0	0	0	0	0
8	0	0	0	0	0	0	0	0
10	0	0	0	0	0	0	0	0
15	NA	0	0	0	0	0	0	0
20	NA	NA	123	190	249	184	0	0
30	NA	NA	NA	334	398	393	334	0
50	NA	NA	NA	NA	NA	714	707	579
100	NA	NA	NA	NA	NA	NA	NA	1,600
27 to 36°F	Local 99% Winter Design Temperature: 27 to 36°F							
6	0	0	68	116	156	180	212	266
8	0	0	82	127	167	187	214	263
10	0	51	97	141	183	210	225	265
15	NA	111	142	183	233	253	274	305
20	NA	NA	187	230	284	307	330	362
30	NA	NA	NA	330	319	419	445	485
50	NA	NA	NA	NA	NA	672	705	763
100	NA	NA	NA	NA	NA	NA	NA	1,554

TABLE 504.3(7b)
EXTERIOR MASONRY CHIMNEY—continued

Minimum Allowable Input Rating of Space-Heating Appliance in Thousands of Btu per Hour

VENT HEIGHT (feet)	INTERNAL AREA OF CHIMNEY (square inches)							
	12	19	28	38	50	63	78	113
17 to 26°F	Local 99% Winter Design Temperature: 17 to 26°F							
6	0	55	99	141	182	215	259	349
8	52	74	111	154	197	226	264	352
10	NA	90	125	169	214	245	278	358
15	NA	NA	167	212	263	296	331	398
20	NA	NA	212	258	316	352	387	457
30	NA	NA	NA	362	429	470	507	581
50	NA	NA	NA	NA	NA	723	766	862
100	NA	NA	NA	NA	NA	NA	NA	1,669
5 to 16°F	Local 99% Winter Design Temperature: 5 to 16°F							
6	NA	78	121	166	214	252	301	416
8	NA	94	135	182	230	269	312	423
10	NA	111	149	198	250	289	331	430
15	NA	NA	193	247	305	346	393	485
20	NA	NA	NA	293	360	408	450	547
30	NA	NA	NA	377	450	531	580	682
50	NA	NA	NA	NA	NA	797	853	972
100	NA	NA	NA	NA	NA	NA	NA	1,833
-10 to 4°F	Local 99% Winter Design Temperature: -10 to 4°F							
6	NA	NA	145	196	249	296	349	484
8	NA	NA	159	213	269	320	371	494
10	NA	NA	175	231	292	339	397	513
15	NA	NA	NA	283	351	404	457	586
20	NA	NA	NA	333	408	468	528	650
30	NA	NA	NA	NA	NA	603	667	805
50	NA	NA	NA	NA	NA	NA	955	1,003
100	NA	NA	NA	NA	NA	NA	NA	NA
-11°F or Lower	Local 99% Winter Design Temperature: -11°F or Lower							
	Not recommended for any vent configurations							

Note: See Figure B-19 in Appendix B for a map showing local 99 percent winter design temperatures in the United States.

For SI: °C = [(°F - 32]/1.8, 1 inch = 25.4 mm, 1 square inch = 645.16 mm², 1 foot = 304.8 mm, 1 British thermal unit per hour = 0.2931 W.

CHAPTER 6

SPECIFIC APPLIANCES

SECTION 601 (IFGC)
GENERAL

601.1 Scope. This chapter shall govern the approval, design, installation, construction, maintenance, alteration and repair of the appliances and equipment specifically identified herein.

SECTION 602 (IFGC)
DECORATIVE APPLIANCES
FOR INSTALLATION IN FIREPLACES

602.1 General. Decorative appliances for installation in approved solid fuel-burning fireplaces shall be tested in accordance with ANSI Z21.60 and shall be installed in accordance with the manufacturer's installation instructions. Manually lighted natural gas decorative appliances shall be tested in accordance with ANSI Z21.84.

602.2 Flame safeguard device. Decorative appliances for installation in approved solid fuel-burning fireplaces, with the exception of those tested in accordance with ANSI Z21.84, shall utilize a direct ignition device, an ignitor or a pilot flame to ignite the fuel at the main burner, and shall be equipped with a flame safeguard device. The flame safeguard device shall automatically shut off the fuel supply to a main burner or group of burners when the means of ignition of such burners becomes inoperative.

602.3 Prohibited installations. Decorative appliances for installation in fireplaces shall not be installed where prohibited by Section 303.3.

SECTION 603 (IFGC)
LOG LIGHTERS

603.1 General. Log lighters shall be tested in accordance with CSA 8 and installed in accordance with the manufacturer's installation instructions.

SECTION 604 (IFGC)
VENTED GAS FIREPLACES
(DECORATIVE APPLIANCES)

604.1 General. Vented gas fireplaces shall be tested in accordance with ANSI Z21.50, shall be installed in accordance with the manufacturer's installation instructions and shall be designed and equipped as specified in Section 602.2.

604.2 Access. Panels, grilles and access doors that are required to be removed for normal servicing operations shall not be attached to the building.

SECTION 605 (IFGC)
VENTED GAS FIREPLACE HEATERS

605.1 General. Vented gas fireplace heaters shall be installed in accordance with the manufacturer's installation instructions, shall be tested in accordance with ANSI Z21.88 and shall be designed and equipped as specified in Section 602.2.

SECTION 606 (IFGC)
INCINERATORS AND CREMATORIES

606.1 General. Incinerators and crematories shall be installed in accordance with the manufacturer's installation instructions.

SECTION 607 (IFGC)
COMMERCIAL-INDUSTRIAL INCINERATORS

607.1 Incinerators, commercial-industrial. Commercial-industrial-type incinerators shall be constructed and installed in accordance with NFPA 82.

SECTION 608 (IFGC)
VENTED WALL FURNACES

608.1 General. Vented wall furnaces shall be tested in accordance with ANSI Z21.86/CSA 2.32 and shall be installed in accordance with the manufacturer's installation instructions.

608.2 Venting. Vented wall furnaces shall be vented in accordance with Section 503.

608.3 Location. Vented wall furnaces shall be located so as not to cause a fire hazard to walls, floors, combustible furnishings or doors. Vented wall furnaces installed between bathrooms and adjoining rooms shall not circulate air from bathrooms to other parts of the building.

608.4 Door swing. Vented wall furnaces shall be located so that a door cannot swing within 12 inches (305 mm) of an air inlet or air outlet of such furnace measured at right angles to the opening. Doorstops or door closers shall not be installed to obtain this clearance.

608.5 Ducts prohibited. Ducts shall not be attached to wall furnaces. Casing extension boots shall not be installed unless listed as part of the appliance.

608.6 Access. Vented wall furnaces shall be provided with access for cleaning of heating surfaces, removal of burners, replacement of sections, motors, controls, filters and other working parts, and for adjustments and lubrication of parts requiring such attention. Panels, grilles and access doors that are required to be removed for normal servicing operations shall not be attached to the building construction.

SECTION 609 (IFGC)
FLOOR FURNACES

609.1 General. Floor furnaces shall be tested in accordance with ANSI Z21.86/CSA 2.32 and shall be installed in accordance with the manufacturer's installation instructions.

609.2 Placement. The following provisions apply to floor furnaces:

1. Floors. Floor furnaces shall not be installed in the floor of any doorway, stairway landing, aisle or passageway of any enclosure, public or private, or in an exitway from any such room or space.

2. Walls and corners. The register of a floor furnace with a horizontal warm-air outlet shall not be placed closer than 6 inches (152 mm) to the nearest wall. A distance of at least 18 inches (457 mm) from two adjoining sides of the floor furnace register to walls shall be provided to eliminate the necessity of occupants walking over the warm-air discharge. The remaining sides shall be permitted to be placed not closer than 6 inches (152 mm) to a wall. Wall-register models shall not be placed closer than 6 inches (152 mm) to a corner.

3. Draperies. The furnace shall be placed so that a door, drapery or similar object cannot be nearer than 12 inches (305 mm) to any portion of the register of the furnace.

4. Floor construction. Floor furnaces shall not be installed in concrete floor construction built on grade.

5. Thermostat. The controlling thermostat for a floor furnace shall be located within the same room or space as the floor furnace or shall be located in an adjacent room or space that is permanently open to the room or space containing the floor furnace.

609.3 Bracing. The floor around the furnace shall be braced and headed with a support framework designed in accordance with the *International Building Code*.

609.4 Clearance. The lowest portion of the floor furnace shall have not less than a 6-inch (152 mm) clearance from the grade level; except where the lower 6-inch (152 mm) portion of the floor furnace is sealed by the manufacturer to prevent entrance of water, the minimum clearance shall be not less than 2 inches (51 mm). Where such clearances cannot be provided, the ground below and to the sides shall be excavated to form a pit under the furnace so that the required clearance is provided beneath the lowest portion of the furnace. A 12-inch (305 mm) minimum clearance shall be provided on all sides except the control side, which shall have an 18-inch (457 mm) minimum clearance.

609.5 First floor installation. Where the basement story level below the floor in which a floor furnace is installed is utilized as habitable space, such floor furnaces shall be enclosed as specified in Section 609.6 and shall project into a nonhabitable space.

609.6 Upper floor installations. Floor furnaces installed in upper stories of buildings shall project below into nonhabitable space and shall be separated from the nonhabitable space by an enclosure constructed of noncombustible materials. The floor furnace shall be provided with access, clearance to all sides and

bottom of not less than 6 inches (152 mm) and combustion air in accordance with Section 304.

SECTION 610 (IFGC)
DUCT FURNACES

610.1 General. Duct furnaces shall be tested in accordance with ANSI Z83.8 or UL 795 and shall be installed in accordance with the manufacturer's installation instructions.

610.2 Access panels. Ducts connected to duct furnaces shall have removable access panels on both the upstream and downstream sides of the furnace.

610.3 Location of draft hood and controls. The controls, combustion air inlets and draft hoods for duct furnaces shall be located outside of the ducts. The draft hood shall be located in the same enclosure from which combustion air is taken.

610.4 Circulating air. Where a duct furnace is installed so that supply ducts convey air to areas outside the space containing the furnace, the return air shall also be conveyed by a duct(s) sealed to the furnace casing and terminating outside the space containing the furnace.

The duct furnace shall be installed on the positive pressure side of the circulating air blower.

SECTION 611 (IFGC)
NONRECIRCULATING DIRECT-FIRED
INDUSTRIAL AIR HEATERS

611.1 General. Nonrecirculating direct-fired industrial air heaters shall be listed to ANSI Z83.4/CSA 3.7 and shall be installed in accordance with the manufacturer's instructions.

611.2 Installation. Nonrecirculating direct-fired industrial air heaters shall not be used to supply any area containing sleeping quarters. Nonrecirculating direct-fired industrial air heaters shall be installed only in industrial or commercial occupancies. Nonrecirculating direct-fired industrial air heaters shall be permitted to provide ventilation air.

611.3 Clearance from combustible materials. Nonrecirculating direct-fired industrial air heaters shall be installed with a clearance from combustible materials of not less than that shown on the rating plate and in the manufacturer's instructions.

611.4 Supply air. All air handled by a nonrecirculating direct-fired industrial air heater, including combustion air, shall be ducted directly from the outdoors.

611.5 Outdoor air louvers. If outdoor air louvers of either the manual or automatic type are used, such devices shall be proven to be in the open position prior to allowing the main burners to operate.

611.6 Atmospheric vents and gas reliefs or bleeds. Nonrecirculating direct-fired industrial air heaters with valve train components equipped with atmospheric vents or gas reliefs or bleeds shall have their atmospheric vent lines or gas reliefs or bleeds lead to the outdoors. Means shall be employed on these lines to prevent water from entering and to prevent blockage by insects and foreign matter. An atmospheric vent line shall not be required to be provided on a valve train component equipped with a listed vent limiter.

611.7 Relief opening. The design of the installation shall include provisions to permit nonrecirculating direct-fired industrial air heaters to operate at rated capacity without overpressurizing the space served by the heaters by taking into account the structure's designed infiltration rate, providing properly designed relief openings or an interlocked power exhaust system, or a combination of these methods. The structure's designed infiltration rate and the size of relief openings shall be determined by approved engineering methods. Relief openings shall be permitted to be louvers or counterbalanced gravity dampers. Motorized dampers or closable louvers shall be permitted to be used, provided they are verified to be in their full open position prior to main burner operation.

611.8 Access. Nonrecirculating direct-fired industrial air heaters shall be provided with access for removal of burners; replacement of motors, controls, filters and other working parts; and for adjustment and lubrication of parts requiring maintenance.

611.9 Purging. Inlet ducting, where used, shall be purged by not less than four air changes prior to an ignition attempt.

SECTION 612 (IFGC)
RECIRCULATING DIRECT-FIRED
INDUSTRIAL AIR HEATERS

612.1 General. Recirculating direct-fired industrial air heaters shall be listed to ANSI Z83.18 and shall be installed in accordance with the manufacturer's installation instructions.

612.2 Location. Recirculating direct-fired industrial air heaters shall be installed only in industrial and commercial occupancies. Recirculating direct-fired air heaters shall not serve any area containing sleeping quarters. Recirculating direct-fired industrial air heaters shall not be installed in hazardous locations or in buildings that contain flammable solids, liquids or gases, explosive materials or substances that can become toxic when exposed to flame or heat.

612.3 Installation. Direct-fired industrial air heaters shall be permitted to be installed in accordance with their listing and the manufacturer's instructions. Direct-fired industrial air heaters shall be installed only in industrial or commercial occupancies. Direct-fired industrial air heaters shall be permitted to provide fresh air ventilation.

612.4 Clearance from combustible materials. Direct-fired industrial air heaters shall be installed with a clearance from combustible material of not less than that shown on the label and in the manufacturer's instructions.

612.5 Air supply. Air to direct-fired industrial air heaters shall be taken from the building, ducted directly from outdoors, or a combination of both. Direct-fired industrial air heaters shall incorporate a means to supply outside ventilation air to the space at a rate of not less than 4 cubic feet per minute per 1,000 Btu per hour (0.38 m³ per min per kW) of rated input of the heater. If a separate means is used to supply ventilation air, an interlock shall be provided so as to lock out the main burner operation until the mechanical means is verified. Where out-side air dampers or closing louvers are used, they shall be verified to be in the open position prior to main burner operation.

612.6 Atmospheric vents, gas reliefs or bleeds. Direct-fired industrial air heaters with valve train components equipped with atmospheric vents, gas reliefs or bleeds shall have their atmospheric vent lines and gas reliefs or bleeds lead to the outdoors.

Means shall be employed on these lines to prevent water from entering and to prevent blockage by insects and foreign matter. An atmospheric vent line shall not be required to be provided on a valve train component equipped with a listed vent limiter.

612.7 Relief opening. The design of the installation shall include adequate provision to permit direct-fired industrial air heaters to operate at rated capacity by taking into account the structure's designed infiltration rate, providing properly designed relief openings or an interlocked power exhaust system, or a combination of these methods. The structure's designed infiltration rate and the size of relief openings shall be determined by approved engineering methods. Relief openings shall be permitted to be louvers or counterbalanced gravity dampers. Motorized dampers or closable louvers shall be permitted to be used, provided they are verified to be in their full open position prior to main burner operation.

SECTION 613 (IFGC)
CLOTHES DRYERS

613.1 General. Clothes dryers shall be tested in accordance with ANSI Z21.5.1 or ANSI Z21.5.2 and shall be installed in accordance with the manufacturer's installation instructions.

SECTION 614 (IFGC)
CLOTHES DRYER EXHAUST

[M] 614.1 Installation. Clothes dryers shall be exhausted in accordance with the manufacturer's instructions. Dryer exhaust systems shall be independent of all other systems, shall convey the moisture and any products of combustion to the outside of the building.

[M] 614.2 Duct penetrations. Ducts that exhaust clothes dryers shall not penetrate or be located within any fireblocking, draftstopping or any wall, floor/ceiling or other assembly required by the *International Building Code* to be fire-resistance rated, unless such duct is constructed of galvanized steel or aluminum of the thickness specified in Table 603.4 of the *International Mechanical Code* and the fire-resistance rating is maintained in accordance with the *International Building Code*. Fire dampers shall not be installed in clothes dryer exhaust duct systems.

[M] 614.3 Cleaning access. Each vertical duct riser for dryers listed to ANSI Z21.5.2 shall be provided with a cleanout or other means for cleaning the interior of the duct.

[M] 614.4 Exhaust installation. Exhaust ducts for clothes dryers shall terminate on the outside of the building and shall be equipped with a backdraft damper. Screens shall not be installed at the duct termination. Ducts shall not be connected or installed with sheet metal screws or other fasteners that will obstruct the flow. Clothes dryer exhaust ducts shall not be connected to a vent connector, vent or chimney. Clothes dryer exhaust ducts shall not extend into or through ducts or plenums.

[M] 614.5 Makeup air. Installations exhausting more than 200 cfm (0.09 m³/s) shall be provided with makeup air. Where a closet is designed for the installation of a clothes dryer, an opening having an area of not less than 100 square inches (645 mm²) for makeup air shall be provided in the closet enclosure, or makeup air shall be provided by other approved means.

[M] 614.6 Domestic clothes dryer ducts. Exhaust ducts for domestic clothes dryers shall be constructed of metal and shall have a smooth interior finish. The exhaust duct shall be a minimum nominal size of 4 inches (102 mm) in diameter. The entire exhaust system shall be supported and secured in place. The male end of the duct at overlapped duct joints shall extend in the direction of airflow. Clothes dryer transition ducts used to connect the appliance to the exhaust duct system shall be metal and limited to a single length not to exceed 8 feet (2438 mm) and shall be listed and labeled for the application. Transition ducts shall not be concealed within construction.

[M] 614.6.1 Maximum length. The maximum length of a clothes dryer exhaust duct shall not exceed 25 feet (7620 mm) from the dryer location to the outlet terminal. The maximum length of the duct shall be reduced 2¹/₂ feet (762 mm) for each 45-degree (0.79 rad) bend and 5 feet (1524 mm) for each 90-degree (1.57 rad) bend. The maximum length of the exhaust duct does not include the transition duct.

Exception: Where the make and model of the clothes dryer to be installed is known and the manufacturer's installation instructions for such dryer are provided to the code official, the maximum length of the exhaust duct, including any transition duct, shall be permitted to be in accordance with the dryer manufacturer's installation instructions.

[M] 614.6.2 Rough-in required. Where a compartment or space for a domestic clothes dryer is provided, an exhaust duct system shall be installed.

[M] 614.7 Commercial clothes dryers. The installation of dryer exhaust ducts serving Type 2 clothes dryers shall comply with the appliance manufacturer's installation instructions. Exhaust fan motors installed in exhaust systems shall be located outside of the airstream. In multiple installations, the fan shall operate continuously or be interlocked to operate when any individual unit is operating. Ducts shall have a minimum clearance of 6 inches (152 mm) to combustible materials.

SECTION 615 (IFGC)
SAUNA HEATERS

615.1 General. Sauna heaters shall be installed in accordance with the manufacturer's installation instructions.

615.2 Location and protection. Sauna heaters shall be located so as to minimize the possibility of accidental contact by a person in the room.

615.2.1 Guards. Sauna heaters shall be protected from accidental contact by an approved guard or barrier of material having a low coefficient of thermal conductivity. The guard shall not substantially affect the transfer of heat from the heater to the room.

615.3 Access. Panels, grilles and access doors that are required to be removed for normal servicing operations shall not be attached to the building.

615.4 Combustion and dilution air intakes. Sauna heaters of other than the direct-vent type shall be installed with the draft hood and combustion air intake located outside the sauna room. Where the combustion air inlet and the draft hood are in a dressing room adjacent to the sauna room, there shall be provisions to prevent physically blocking the combustion air inlet and the draft hood inlet, and to prevent physical contact with the draft hood and vent assembly, or warning notices shall be posted to avoid such contact. Any warning notice shall be easily readable, shall contrast with its background and the wording shall be in letters not less than ¹/₄ inch (6.4 mm) high.

615.5 Combustion and ventilation air. Combustion air shall not be taken from inside the sauna room. Combustion and ventilation air for a sauna heater not of the direct-vent type shall be provided to the area in which the combustion air inlet and draft hood are located in accordance with Section 304.

615.6 Heat and time controls. Sauna heaters shall be equipped with a thermostat which will limit room temperature to 194°F (90°C). If the thermostat is not an integral part of the sauna heater, the heat-sensing element shall be located within 6 inches (152 mm) of the ceiling. If the heat-sensing element is a capillary tube and bulb, the assembly shall be attached to the wall or other support, and shall be protected against physical damage.

615.6.1 Timers. A timer, if provided to control main burner operation, shall have a maximum operating time of 1 hour. The control for the timer shall be located outside the sauna room.

615.7 Sauna room. A ventilation opening into the sauna room shall be provided. The opening shall be not less than 4 inches by 8 inches (102 mm by 203 mm) located near the top of the door into the sauna room.

615.7.1 Warning notice. The following permanent notice, constructed of approved material, shall be mechanically attached to the sauna room on the outside:

WARNING: DO NOT EXCEED 30 MINUTES IN SAUNA. EXCESSIVE EXPOSURE CAN BE HARMFUL TO HEALTH. ANY PERSON WITH POOR HEALTH SHOULD CONSULT A PHYSICIAN BEFORE USING SAUNA.

The words shall contrast with the background and the wording shall be in letters not less than ¹/₄ inch (6.4 mm) high.

Exception: This section shall not apply to one- and two-family dwellings.

SECTION 616 (IFGC)
ENGINE AND GAS TURBINE-POWERED EQUIPMENT

616.1 Powered equipment. Permanently installed equipment powered by internal combustion engines and turbines shall be installed in accordance with the manufacturer's installation instructions and NFPA 37.

SECTION 617 (IFGC)
POOL AND SPA HEATERS

617.1 General. Pool and spa heaters shall be tested in accordance with ANSI Z21.56 and shall be installed in accordance with the manufacturer's installation instructions.

SECTION 618 (IFGC)
FORCED-AIR WARM-AIR FURNACES

618.1 General. Forced-air warm-air furnaces shall be tested in accordance with ANSI Z21.47 or UL 795 and shall be installed in accordance with the manufacturer's installation instructions.

618.2 Forced-air furnaces. The minimum unobstructed total area of the outside and return air ducts or openings to a forced-air warm-air furnace shall be not less than 2 square inches for each 1,000 Btu/h (4402 mm²/W) output rating capacity of the furnace and not less than that specified in the furnace manufacturer's installation instructions. The minimum unobstructed total area of supply ducts from a forced-air warm-air furnace shall be not less than 2 square inches for each 1,000 Btu/h (4402 mm²/W) output rating capacity of the furnace and not less than that specified in the furnace manufacturer's installation instructions.

Exception: The total area of the supply air ducts and outside and return air ducts shall not be required to be larger than the minimum size required by the furnace manufacturer's installation instructions.

618.3 Dampers. Volume dampers shall not be placed in the air inlet to a furnace in a manner that will reduce the required air to the furnace.

618.4 Circulating air ducts for forced-air warm-air furnaces. Circulating air for fuel-burning, forced-air-type, warm-air furnaces shall be conducted into the blower housing from outside the furnace enclosure by continuous air-tight ducts.

618.5 Prohibited sources. Outside or return air for a forced-air heating system shall not be taken from the following locations:

1. Closer than 10 feet (3048 mm) from an appliance vent outlet, a vent opening from a plumbing drainage system or the discharge outlet of an exhaust fan, unless the outlet is 3 feet (914 mm) above the outside air inlet.

2. Where there is the presence of objectionable odors, fumes or flammable vapors; or where located less than 10 feet (3048 mm) above the surface of any abutting public way or driveway; or where located at grade level by a sidewalk, street, alley or driveway.

3. A hazardous or insanitary location or a refrigeration machinery room as defined in the *International Mechanical Code.*

4. A room or space, the volume of which is less than 25 percent of the entire volume served by such system. Where connected by a permanent opening having an area sized in accordance with Section 618.2, adjoining rooms or spaces shall be considered as a single room or space for the purpose of determining the volume of such rooms or spaces.

   Exception: The minimum volume requirement shall not apply where the amount of return air taken from a room or space is less than or equal to the amount of supply air delivered to such room or space.

5. A room or space containing an appliance where such a room or space serves as the sole source of return air.

   Exception: This shall not apply where:

   1. The appliance is a direct-vent appliance or an appliance not requiring a vent in accordance with Section 501.8.

   2. The room or space complies with the following requirements:

      2.1. The return air shall be taken from a room or space having a volume exceeding 1 cubic foot for each 10 Btu/h (9.6 L/W) of combined input rating of all fuel-burning appliances therein.

      2.2. The volume of supply air discharged back into the same space shall be approximately equal to the volume of return air taken from the space.

      2.3. Return-air inlets shall not be located within 10 feet (3048 mm) of any appliance firebox or draft hood in the same room or space.

   3. Rooms or spaces containing solid fuel-burning appliances, provided that return-air inlets are located not less than 10 feet (3048 mm) from the firebox of such appliances.

6. A closet, bathroom, toilet room, kitchen, garage, mechanical room, boiler room or furnace room.

618.6 Screen. Required outdoor air inlets for residential portions of a building shall be covered with a screen having ¹/₄-inch (6.4 mm) openings. Required outdoor air inlets serving a nonresidential portion of a building shall be covered with screen having openings larger than ¹/₄ inch (6.4 mm) and not larger than 1 inch (25 mm).

618.7 Return-air limitation. Return air from one dwelling unit shall not be discharged into another dwelling unit.

SECTION 619 (IFGC)
CONVERSION BURNERS

619.1 Conversion burners. The installation of conversion burners shall conform to ANSI Z21.8.

SECTION 620 (IFGC)
UNIT HEATERS

620.1 General. Unit heaters shall be tested in accordance with ANSI Z83.8 and shall be installed in accordance with the manufacturer's installation instructions.

620.2 Support. Suspended-type unit heaters shall be supported by elements that are designed and constructed to accommodate the weight and dynamic loads. Hangers and brackets shall be of noncombustible material.

620.3 Ductwork. Ducts shall not be connected to a unit heater unless the heater is listed for such installation.

620.4 Clearance. Suspended-type unit heaters shall be installed with clearances to combustible materials of not less than 18 inches (457 mm) at the sides, 12 inches (305 mm) at the bottom and 6 inches (152 mm) above the top where the unit heater has an internal draft hood or 1 inch (25 mm) above the top of the sloping side of the vertical draft hood.

Floor-mounted-type unit heaters shall be installed with clearances to combustible materials at the back and one side only of not less than 6 inches (152 mm). Where the flue gases are vented horizontally, the 6-inch (152 mm) clearance shall be measured from the draft hood or vent instead of the rear wall of the unit heater. Floor-mounted-type unit heaters shall not be installed on combustible floors unless listed for such installation.

Clearances for servicing all unit heaters shall be in accordance with the manufacturer's installation instructions.

Exception: Unit heaters listed for reduced clearance shall be permitted to be installed with such clearances in accordance with their listing and the manufacturer's instructions.

SECTION 621 (IFGC)
UNVENTED ROOM HEATERS

621.1 General. Unvented room heaters shall be tested in accordance with ANSI Z21.11.2 and shall be installed in accordance with the conditions of the listing and the manufacturer's installation instructions. Unvented room heaters utilizing fuels other than fuel gas shall be regulated by the *International Mechanical Code*.

621.2 Prohibited use. One or more unvented room heaters shall not be used as the sole source of comfort heating in a dwelling unit.

621.3 Input rating. Unvented room heaters shall not have an input rating in excess of 40,000 Btu/h (11.7 Kw).

621.4 Prohibited locations. Unvented room heaters shall not be installed within occupancies in Groups A, E and I. The location of unvented room heaters shall also comply with Section 303.3.

621.5 Room or space volume. The aggregate input rating of all unvented appliances installed in a room or space shall not exceed 20 Btu/h per cubic foot (207 W/m³) of volume of such room or space. Where the room or space in which the equipment is installed is directly connected to another room or space by a doorway, archway or other opening of comparable size

that cannot be closed, the volume of such adjacent room or space shall be permitted to be included in the calculations.

621.6 Oxygen-depletion safety system. Unvented room heaters shall be equipped with an oxygen-depletion-sensitive safety shutoff system. The system shall shut off the gas supply to the main and pilot burners when the oxygen in the surrounding atmosphere is depleted to the percent concentration specified by the manufacturer, but not lower than 18 percent. The system shall not incorporate field adjustment means capable of changing the set point at which the system acts to shut off the gas supply to the room heater.

621.7 Unvented decorative room heaters. An unvented decorative room heater shall not be installed in a factory-built fireplace unless the fireplace system has been specifically tested, listed and labeled for such use in accordance with UL 127.

621.7.1 Ventless firebox enclosures. Ventless firebox enclosures used with unvented decorative room heaters shall be listed as complying with ANSI Z21.91.

SECTION 622 (IFGC)
VENTED ROOM HEATERS

622.1 General. Vented room heaters shall be tested in accordance with ANSI Z21.86/CSA 2.32, shall be designed and equipped as specified in Section 602.2 and shall be installed in accordance with the manufacturer's installation instructions.

SECTION 623 (IFGC)
COOKING APPLIANCES

623.1 Cooking appliances. Cooking appliances that are designed for permanent installation, including ranges, ovens, stoves, broilers, grills, fryers, griddles, hot plates and barbecues, shall be tested in accordance with ANSI Z21.1, ANSI Z21.58 or ANSI Z83.11 and shall be installed in accordance with the manufacturer's installation instructions.

623.2 Prohibited location. Cooking appliances designed, tested, listed and labeled for use in commercial occupancies shall not be installed within dwelling units or within any area where domestic cooking operations occur.

623.3 Domestic appliances. Cooking appliances installed within dwelling units and within areas where domestic cooking operations occur shall be listed and labeled as household-type appliances for domestic use.

623.4 Domestic range installation. Domestic ranges installed on combustible floors shall be set on their own bases or legs and shall be installed with clearances of not less than that shown on the label.

623.5 Open-top broiler unit hoods. A ventilating hood shall be provided above a domestic open-top broiler unit, unless otherwise listed for forced down draft ventilation.

623.5.1 Clearances. A minimum clearance of 24 inches (610 mm) shall be maintained between the cooking top and combustible material above the hood. The hood shall be at least as wide as the open-top broiler unit and be centered over the unit.

623.6 Commercial cooking appliance venting. Commercial cooking appliances, other than those exempted by Section 501.8, shall be vented by connecting the appliance to a vent or chimney in accordance with this code and the appliance manufacturer's instructions or the appliance shall be vented in accordance with Section 505.1.1.

SECTION 624 (IFGC)
WATER HEATERS

624.1 General. Water heaters shall be tested in accordance with ANSI Z 21.10.1 and ANSI Z 21.10.3 and shall be installed in accordance with the manufacturer's installation instructions Water heaters utilizing fuels other than fuel gas shall be regulated by the *International Mechanical Code.*

   624.1.1 Installation requirements. The requirements for water heaters relative to sizing, relief valves, drain pans and scald protection shall be in accordance with the *International Plumbing Code.*

624.2 Water heaters utilized for space heating. Water heaters utilized both to supply potable hot water and provide hot water for space-heating applications shall be listed and labeled for such applications by the manufacturer and shall be installed in accordance with the manufacturer's installation instructions and the *International Plumbing Code.*

SECTION 625 (IFGC)
REFRIGERATORS

625.1 General. Refrigerators shall be tested in accordance with ANSI Z21.19 and shall be installed in accordance with the manufacturer's installation instructions.

Refrigerators shall be provided with adequate clearances for ventilation at the top and back, and shall be installed in accordance with the manufacturer's instructions. If such instructions are not available, at least 2 inches (51 mm) shall be provided between the back of the refrigerator and the wall and at least 12 inches (305 mm) above the top.

SECTION 626 (IFGC)
GAS-FIRED TOILETS

626.1 General. Gas-fired toilets shall be tested in accordance with ANSI Z21.61 and installed in accordance with the manufacturer's installation instructions.

626.2 Clearance. A gas-fired toilet shall be installed in accordance with its listing and the manufacturer's instructions, provided that the clearance shall in any case be sufficient to afford ready access for use, cleanout and necessary servicing.

SECTION 627 (IFGC)
AIR-CONDITIONING EQUIPMENT

627.1 General. Gas-fired air-conditioning equipment shall be tested in accordance with ANSI Z21.40.1 or ANSI Z21.40.2 and shall be installed in accordance with the manufacturer's installation instructions.

627.2 Independent piping. Gas piping serving heating equipment shall be permitted to also serve cooling equipment where such heating and cooling equipment cannot be operated simultaneously (see Section 402).

627.3 Connection of gas engine-powered air conditioners. To protect against the effects of normal vibration in service, gas engines shall not be rigidly connected to the gas supply piping.

627.4 Clearances for indoor installation. Air-conditioning equipment installed in rooms other than alcoves and closets shall be installed with clearances not less than those specified in Section 308.3 except that air-conditioning equipment listed for installation at lesser clearances than those specified in Section 308.3 shall be permitted to be installed in accordance with such listing and the manufacturer's instructions and air-conditioning equipment listed for installation at greater clearances than those specified in Section 308.3 shall be installed in accordance with such listing and the manufacturer's instructions.

Air-conditioning equipment installed in rooms other than alcoves and closets shall be permitted to be installed with reduced clearances to combustible material, provided that the combustible material is protected in accordance with Table 308.2.

627.5 Alcove and closet installation. Air-conditioning equipment installed in spaces such as alcoves and closets shall be specifically listed for such installation and installed in accordance with the terms of such listing. The installation clearances for air-conditioning equipment in alcoves and closets shall not be reduced by the protection methods described in Table 308.2.

627.6 Installation. Air-conditioning equipment shall be installed in accordance with the manufacturer's instructions. Unless the equipment is listed for installation on a combustible surface such as a floor or roof, or unless the surface is protected in an approved manner, equipment shall be installed on a surface of noncombustible construction with noncombustible material and surface finish and with no combustible material against the underside thereof.

627.7 Plenums and air ducts. A plenum supplied as a part of the air-conditioning equipment shall be installed in accordance with the equipment manufacturer's instructions. Where a plenum is not supplied with the equipment, such plenum shall be installed in accordance with the fabrication and installation instructions provided by the plenum and equipment manufacturer. The method of connecting supply and return ducts shall facilitate proper circulation of air.

Where air-conditioning equipment is installed within a space separated from the spaces served by the equipment, the air circulated by the equipment shall be conveyed by ducts that are sealed to the casing of the equipment and that separate the circulating air from the combustion and ventilation air.

627.8 Refrigeration coils. A refrigeration coil shall not be installed in conjunction with a forced-air furnace where circulation of cooled air is provided by the furnace blower, unless the blower has sufficient capacity to overcome the external static resistance imposed by the duct system and cooling coil at the air throughput necessary for heating or cooling, whichever is greater. Furnaces shall not be located upstream from cooling units, unless the cooling unit is designed or equipped so as not to develop excessive temperature or pressure. Refrigeration coils shall be installed in parallel with or on the downstream side of central furnaces to avoid condensation in the heating element, unless the furnace has been specifically listed for downstream installation. With a parallel flow arrangement, the dampers or other means used to control flow of air shall be sufficiently tight to prevent any circulation of cooled air through the furnace.

Means shall be provided for disposal of condensate and to prevent dripping of condensate onto the heating element.

627.9 Cooling units used with heating boilers. Boilers, where used in conjunction with refrigeration systems, shall be installed so that the chilled medium is piped in parallel with the heating boiler with appropriate valves to prevent the chilled medium from entering the heating boiler. Where hot water heating boilers are connected to heating coils located in air-handling units where they might be exposed to refrigerated air circulation, such boiler piping systems shall be equipped with flow control valves or other automatic means to prevent gravity circulation of the boiler water during the cooling cycle.

627.10 Switches in electrical supply line. Means for interrupting the electrical supply to the air-conditioning equipment and to its associated cooling tower (if supplied and installed in a location remote from the air conditioner) shall be provided within sight of and not over 50 feet (15 240 mm) from the air conditioner and cooling tower.

SECTION 628 (IFGC)
ILLUMINATING APPLIANCES

628.1 General. Illuminating appliances shall be tested in accordance with ANSI Z21.42 and shall be installed in accordance with the manufacturer's installation instructions.

628.2 Mounting on buildings. Illuminating appliances designed for wall or ceiling mounting shall be securely attached to substantial structures in such a manner that they are not dependent on the gas piping for support.

628.3 Mounting on posts. Illuminating appliances designed for post mounting shall be securely and rigidly attached to a post. Posts shall be rigidly mounted. The strength and rigidity of posts greater than 3 feet (914 mm) in height shall be at least equivalent to that of a $2^1/_2$-inch-diameter (64 mm) post constructed of 0.064-inch-thick (1.6-mm) steel or a 1-inch (25.4 mm) Schedule 40 steel pipe. Posts 3 feet (914 mm) or less in height shall not be smaller than a $^3/_4$-inch (19.1 mm) Schedule 40 steel pipe. Drain openings shall be provided near the base of posts where there is a possibility of water collecting inside them.

628.4 Appliance pressure regulators. Where an appliance pressure regulator is not supplied with an illuminating appliance and the service line is not equipped with a service pressure regulator, an appliance pressure regulator shall be installed in the line to the illuminating appliance. For multiple installations, one regulator of adequate capacity shall be permitted to serve more than one illuminating appliance.

SECTION 629 (IFGC)
SMALL CERAMIC KILNS

629.1 General. Ceramic kilns with a maximum interior volume of 20 cubic feet (0.566 m³) and used for hobby and noncommercial purposes shall be installed in accordance with the manufacturer's installation instructions and the provisions of this code.

SECTION 630 (IFGC)
INFRARED RADIANT HEATERS

630.1 General. Infrared radiant heaters shall be tested in accordance with ANSI Z83.6 and shall be installed in accordance with the manufacturer's installation instructions.

630.2 Support. Infrared radiant heaters shall be fixed in a position independent of gas and electric supply lines. Hangers and brackets shall be of noncombustible material.

SECTION 631 (IFGC)
BOILERS

631.1 Standards. Boilers shall be listed in accordance with the requirements of ANSI Z21.13 or UL 795. If applicable, the boiler shall be designed and constructed in accordance with the requirements of ASME CSD-1 and as applicable, the ASME *Boiler and Pressure Vessel Code*, Sections I, II, IV, V and IX and NFPA 85.

631.2 Installation. In addition to the requirements of this code, the installation of boilers shall be in accordance with the manufacturer's instructions and the *International Mechanical Code*. Operating instructions of a permanent type shall be attached to the boiler. Boilers shall have all controls set, adjusted and tested by the installer. A complete control diagram together with complete boiler operating instructions shall be furnished by the installer. The manufacturer's rating data and the nameplate shall be attached to the boiler.

631.3 Clearance to combustible materials. Clearances to combustible materials shall be in accordance with Section 308.4.

SECTION 632 (IFGC)
EQUIPMENT INSTALLED IN EXISTING UNLISTED BOILERS

632.1 General. Gas equipment installed in existing unlisted boilers shall comply with Section 631.1 and shall be installed in accordance with the manufacturer's instructions and the *International Mechanical Code*.

SECTION 633 (IFGC)
STATIONARY FUEL-CELL POWER SYSTEMS

[F] 633.1 General. Stationary fuel-cell power systems having a power output not exceeding 10 MW shall be tested in accordance with ANSI CSA America FC 1 and shall be installed in accordance with the manufacturer's installation instructions and NFPA 853.

SECTION 634 (IFGS)
CHIMNEY DAMPER OPENING AREA

634.1 Free opening area of chimney dampers. Where an unlisted decorative appliance for installation in a vented fireplace is installed, the fireplace damper shall have a permanent free opening equal to or greater than specified in Table 634.1.

SECTION 635 (IFGC)
GASEOUS HYDROGEN SYSTEMS

635.1 Installation. The installation of gaseous hydrogen systems shall be in accordance with the applicable requirements of this code, the *International Fire Code* and the *International Building Code*.

TABLE 634.1
FREE OPENING AREA OF CHIMNEY DAMPER FOR VENTING FLUE GASES
FROM UNLISTED DECORATIVE APPLIANCES FOR INSTALLATION IN VENTED FIREPLACES

CHIMNEY HEIGHT (feet)	MINIMUM PERMANENT FREE OPENING (square inches)[a]						
	8	13	20	29	39	51	64
	Appliance input rating (Btu per hour)						
6	7,800	14,000	23,200	34,000	46,400	62,400	80,000
8	8,400	15,200	25,200	37,000	50,400	68,000	86,000
10	9,000	16,800	27,600	40,400	55,800	74,400	96,400
15	9,800	18,200	30,200	44,600	62,400	84,000	108,800
20	10,600	20,200	32,600	50,400	68,400	94,000	122,200
30	11,200	21,600	36,600	55,200	76,800	105,800	138,600

For SI: 1 inch = 25.4 mm, 1 foot = 304.8 mm, 1 square inch = 645.16 m², 1 British thermal unit per hour = 0.2931 W.

a. The first six minimum permanent free openings (8 to 51 square inches) correspond approximately to the cross-sectional areas of chimneys having diameters of 3 through 8 inches, respectively. The 64-square-inch opening corresponds to the cross-sectional area of standard 8-inch by 8-inch chimney tile.

CHAPTER 7

GASEOUS HYDROGEN SYSTEMS

SECTION 701 (IFGC)
GENERAL

701.1 Scope. The installation of gaseous hydrogen systems shall comply with this chapter and Chapters 30 and 35 of the *International Fire Code*. Compressed gases shall also comply with Chapter 27 of the *International Fire Code* for general requirements.

701.2 Permits. Permits shall be required as set forth in Section 106 and as required by the *International Fire Code*.

SECTION 702 (IFGC)
GENERAL DEFINITIONS

702.1 Definitions. The following words and terms shall, for the purposes of this chapter and as used elsewhere in this code, have the meanings shown herein.

HYDROGEN CUTOFF ROOM. A room or space which is intended exclusively to house a gaseous hydrogen system.

HYDROGEN-GENERATING APPLIANCE. A self-contained package or factory-matched packages of integrated systems for generating gaseous hydrogen. Hydrogen-generating appliances utilize electrolysis, reformation, chemical or other processes to generate hydrogen.

GASEOUS HYDROGEN SYSTEM. An assembly of piping, devices and apparatus designed to generate, store, contain, distribute or transport a nontoxic, gaseous hydrogen containing mixture having at least 95-percent hydrogen gas by volume and not more than 1-percent oxygen by volume. Gaseous hydrogen systems consist of items such as compressed gas containers, reactors and appurtenances, including pressure regulators, pressure relief devices, manifolds, pumps, compressors and interconnecting piping and tubing and controls.

SECTION 703 (IFGC)
GENERAL REQUIREMENTS

703.1 Hydrogen-generating and refueling operations. Ventilation shall be required in accordance with Section 703.1.1, 703.1.2 or 703.1.3 in public garages, private garages, repair garages, automotive motor fuel-dispensing facilities and parking garages that contain hydrogen-generating appliances or refueling systems. For the purpose of this section, rooms or spaces that are not part of the living space of a dwelling unit and that communicate directly with a private garage through openings shall be considered to be part of the private garage.

703.1.1 Natural ventilation. Indoor locations intended for hydrogen-generating or refueling operations shall be limited to a maximum floor area of 850 square feet (79 m²) and shall communicate with the outdoors in accordance with Sections 703.1.1.1 and 703.1.1.2. The maximum rated output capacity of hydrogen generating appliances shall not exceed 4 standard cubic feet per minute (0.00189 m³/s) of hydrogen for each 250 square feet (23.2 m²) of floor area in such spaces. The minimum cross-sectional dimension of air openings shall be 3 inches (76 mm). Where ducts are used, they shall be of the same cross-sectional area as the free area of the openings to which they connect. In such locations, equipment and appliances having an ignition source shall be located such that the source of ignition is not within 12 inches (305 mm) of the ceiling.

703.1.1.1 Two openings. Two permanent openings shall be provided within the garage. The upper opening shall be located entirely within 12 inches (305 mm) of the ceiling of the garage. The lower opening shall be located entirely within 12 inches (305 mm) of the floor of the garage. Both openings shall be provided in the same exterior wall. The openings shall communicate directly with the outdoors and shall have a minimum free area of $^1/_2$ square foot per 1,000 cubic feet (1 m²/610 m³) of garage volume.

703.1.1.2 Louvers and grilles. In calculating the free area required by Section 703.1.1.1, the required size of openings shall be based on the net free area of each opening. If the free area through a design of louver or grille is known, it shall be used in calculating the size opening required to provide the free area specified. If the design and free area are not known, it shall be assumed that wood louvers will have 25-percent free area and metal louvers and grilles will have 75-percent free area. Louvers and grilles shall be fixed in the open position.

703.1.2 Mechanical ventilation. Indoor locations intended for hydrogen-generating or refueling operations shall be ventilated in accordance with Section 502.16 of the *International Mechanical Code*. In such locations, equipment and appliances having an ignition source shall be located such that the source of ignition is below the mechanical ventilation outlet(s).

703.1.3 Specially engineered installations. As an alternative to the provisions of Section 703.1.1 and 703.1.2, the necessary supply of air for ventilation and dilution of flammable gases shall be provided by an approved engineered system.

[F] 703.2 Containers, cylinders and tanks. Compressed gas containers, cylinders and tanks shall comply with Chapters 30 and 35 of the *International Fire Code*.

[F] 703.2.1 Limitations for indoor storage and use. Flammable gas cylinders in occupancies regulated by the *International Residential Code* shall not exceed 250 cubic feet (7.1 m³) at normal temperature and pressure (NTP).

[F] 703.2.2 Design and construction. Compressed gas containers, cylinders and tanks shall be designed, constructed and tested in accordance with the Chapter 27 of the *International Fire Code,* ASME *Boiler and Pressure Vessel Code* (Section VIII) or DOTn 49 CFR, Parts 100-180.

[F] 703.3 Pressure relief devices. Pressure relief devices shall be provided in accordance with Sections 703.3.1 through 703.3.8. Pressure relief devices shall be sized and selected in accordance with CGA S-1.1, CGA S-1.2 and CGA S-1.3.

[F] 703.3.1 Valves between pressure relief devices and containers. Valves including shutoffs, check valves and other mechanical restrictions shall not be installed between the pressure relief device and container being protected by the relief device.

Exception: A locked-open shutoff valve on containers equipped with multiple pressure-relief device installations where the arrangement of the valves provides the full required flow through the minimum number of required relief devices at all times.

[F] 703.3.2 Installation. Valves and other mechanical restrictions shall not be located between the pressure relief device and the point of release to the atmosphere.

[F] 703.3.3 Containers. Containers shall be provided with pressure relief devices in accordance with the ASME *Boiler and Pressure Vessel Code* (Section VIII), DOTn 49 CFR, Parts 100-180 and Section 703.3.7.

[F] 703.3.4 Vessels other than containers. Vessels other than containers shall be protected with pressure relief devices in accordance with the ASME *Boiler and Pressure Vessel Code* (Section VIII), or DOTn 49 CFR, Parts 100-180.

[F] 703.3.5 Sizing. Pressure relief devices shall be sized in accordance with the specifications to which the container was fabricated. The relief device shall be sized to prevent the maximum design pressure of the container or system from being exceeded.

[F] 703.3.6 Protection. Pressure relief devices and any associated vent piping shall be designed, installed and located so that their operation will not be affected by water or other debris accumulating inside the vent or obstructing the vent.

[F] 703.3.7 Access. Pressure relief devices shall be located such that they are provided with ready access for inspection and repair.

[F] 703.3.8 Configuration. Pressure relief devices shall be arranged to discharge unobstructed in accordance with Section 2209 of the *International Fire Code*. Discharge shall be directed to the outdoors in such a manner as to prevent impingement of escaping gas on personnel, containers, equipment and adjacent structures and to prevent introduction of escaping gas into enclosed spaces. The discharge shall not terminate under eaves or canopies.

Exception: This section shall not apply to DOTn-specified containers with an internal volume of 2 cubic feet (0.057 m³) or less.

[F] 703.4 Venting. Relief device vents shall be terminated in an approved location in accordance with Section 2209 of the *International Fire Code*.

[F] 703.5 Security. Compressed gas containers, cylinders, tanks and systems shall be secured against accidental dislodgement in accordance with Chapter 30 of the *International Fire Code*.

[F] 703.6 Electrical wiring and equipment. Electrical wiring and equipment shall comply with the ICC *Electrical Code*.

SECTION 704 (IFGC)
PIPING, USE AND HANDLING

704.1 Applicability. Use and handling of containers, cylinders, tanks and hydrogen gas systems shall comply with this section. Gaseous hydrogen systems, equipment and machinery shall be listed or approved.

704.1.1 Controls. Compressed gas system controls shall be designed to prevent materials from entering or leaving process or reaction systems at other than the intended time, rate or path. Automatic controls shall be designed to be fail safe in accordance with accepted engineering practice.

704.1.2 Piping systems. Piping, tubing, valves and fittings conveying gaseous hydrogen shall be designed and installed in accordance with Sections 704.1.2.1 through 704.1.2.5.1, Chapter 27 of the *International Fire Code*, and ASME B31.3. Cast-iron pipe, valves and fittings shall not be used.

704.1.2.1 Sizing. Gaseous hydrogen piping shall be sized in accordance with approved engineering methods.

704.1.2.2 Identification of hydrogen piping systems. Hydrogen piping systems shall be marked in accordance with ANSI A13.1. Markings used for piping systems shall consist of the name of the contents and shall include a direction-of-flow arrow. Markings shall be provided at all of the following locations:

1. At each valve.

2. At wall, floor and ceiling penetrations.

3. At each change of direction.

4. At intervals not exceeding 20 feet (6096 mm).

704.1.2.3 Piping design and construction. Piping and tubing materials shall be 300 series stainless steel or materials listed or approved for hydrogen service and the use intended through the full range of operating conditions to which they will be subjected. Piping systems shall be designed and constructed to provide allowance for expansion, contraction, vibration, settlement and fire exposure.

704.1.2.3.1 Prohibited locations. Piping shall not be installed in or through a circulating air duct; clothes chute; chimney or gas vent; ventilating duct; dumbwaiter; or elevator shaft. Piping shall not be concealed or covered by the surface of any wall, floor or ceiling.

704.1.2.3.2 Interior piping. Except for through penetrations, piping located inside of buildings shall be installed in exposed locations and provided with ready access for visual inspection.

704.1.2.3.3 Underground piping. Underground piping, including joints and fittings, shall be protected from corrosion and installed in accordance with approved engineered methods.

704.1.2.3.4 Piping through foundation wall. Underground piping shall not penetrate the outer foundation or basement wall of a building.

704.1.2.3.5 Protection against physical damage. In concealed locations, where piping other than stainless steel piping, stainless steel tubing or black steel is installed through holes or notches in wood studs, joists, rafters or similar members less than 1.5 inches (38 mm) from the nearest edge of the member, the pipe shall be protected by shield plates. Shield plates shall be a minimum of $^1/_{16}$-inch-thick (1.6 mm) steel, shall cover the area of the pipe where the member is notched or bored and shall extend a minimum of 4 inches (102 mm) above sole plates, below top plates and to each side of a stud, joist or rafter.

704.1.2.3.6 Piping outdoors. Piping installed above ground, outdoors, shall be securely supported and located where it will be protected from physical damage. Piping passing through an exterior wall of a building shall be encased in a protective pipe sleeve. The annular space between the piping and the sleeve shall be sealed from the inside such that the sleeve is ventilated to the outdoors. Where passing through an exterior wall of a building, the piping shall also be protected against corrosion by coating or wrapping with an inert material. Below-ground piping shall be protected against corrosion.

704.1.2.3.7 Settlement. Piping passing through concrete or masonry walls shall be protected against differential settlement.

704.1.2.4 Joints. Joints in piping and tubing in hydrogen service shall be listed as complying with ASME B31.3 to include the use of welded, brazed, flared, socket, slip and compression fittings. Gaskets and sealants used in hydrogen service shall be listed as complying with ASME B31.3. Threaded and flanged connections shall not be used in areas other than hydrogen cutoff rooms and outdoors.

704.1.2.4.1 Brazed joints. Brazing alloys shall have a melting point greater than 1,000°F (538°C).

704.1.2.4.2 Electrical continuity. Mechanical joints shall maintain electrical continuity through the joint or a bonding jumper shall be installed around the joint.

704.1.2.5 Valves and piping components. Valves, regulators and piping components shall be listed or approved for hydrogen service, shall be provided with access and shall be designed and constructed to withstand the maximum pressure to which such components will be subjected.

704.1.2.5.1 Shutoff valves on storage containers and tanks. Shutoff valves shall be provided on all storage container and tank connections except for pressure relief devices. Shutoff valves shall be provided with ready access.

704.2 Upright use. Compressed gas containers, cylinders and tanks, except those with a water volume less than 1.3 gallons (5 L) and those designed for use in a horizontal position, shall be used in an upright position with the valve end up. An upright position shall include conditions where the container, cylinder or tank axis is inclined as much as 45 degrees (0.79 rad) from the vertical.

704.3 Material-specific regulations. In addition to the requirements of this section, indoor and outdoor use of hydrogen compressed gas shall comply with the material-specific provisions of Chapters 30 and 35 of the *International Fire Code*.

704.4 Handling. The handling of compressed gas containers, cylinders and tanks shall comply with Chapter 27 of the *International Fire Code*.

SECTION 705 (IFGC)
TESTING OF HYDROGEN PIPING SYSTEMS

705.1 General. Prior to acceptance and initial operation, all piping installations shall be inspected and pressure tested to determine that the materials, design fabrication and installation practices comply with the requirements of this code.

705.2 Inspections. Inspections shall consist of a visual examination of the entire piping system installation and a pressure test. Hydrogen piping systems shall be inspected in accordance with this code. Inspection methods such as outlined in ASME B31.3 shall be permitted where specified by the design engineer and approved by the code official. Inspections shall be conducted or verified by the code official prior to system operation.

705.3 Pressure tests. A hydrostatic or pneumatic leak test shall be performed. Testing of hydrogen piping systems shall utilize testing procedures identified in ASME B31.3 or other approved methods, provided that the testing is performed in accordance with the minimum provisions specified in Sections 705.3.1 through 705.4.1.

705.3.1 Hydrostatic leak tests. The hydrostatic test pressure shall be not less than one-and-one-half times the maximum working pressure, and not less than 100 psig (689.5 kPa gauge).

705.3.2 Pneumatic leak tests. The pneumatic test pressure shall be not less than one-and-one-half times the maximum working pressure for systems less than 125 psig (862 kPa gauge) and not less than 5 psig (34.5 kPa gauge), whichever is greater. For working pressures at or above 125 psig (862 kPa gauge), the pneumatic test pressure shall be not less than 110 percent of the maximum working pressure.

705.3.3 Test limits. Where the test pressure exceeds 125 psig (862 kPa gauge), the test pressure shall not exceed a value that produces hoop stress in the piping greater than 50 percent of the specified minimum yield strength of the pipe.

705.3.4 Test medium. Deionized water shall be utilized to perform hydrostatic pressure testing and shall be obtained from a potable source. The medium utilized to perform pneumatic pressure testing shall be air, nitrogen, carbon dioxide or an inert gas; oxygen shall not be used.

705.3.5 Test duration. The minimum test duration shall be $^1/_2$ hour. The test duration shall be not less than $^1/_2$ hour for each 500 cubic feet (14.2 m^3) of pipe volume or fraction thereof. For piping systems having a volume of more than 24,000 cubic feet (680 m^3), the duration of the test shall not be required to exceed 24 hours. The test pressure required in Sections 705.3.1 and 705.3.2 shall be maintained for the entire duration of the test.

705.3.6 Test gauges. Gauges used for testing shall be as follows:

1. Tests requiring a pressure of 10 psig (68.95 kPa gauge) or less shall utilize a testing gauge having increments of 0.10 psi (0.6895 kPa) or less.

2. Tests requiring a pressure greater than 10 psig (68.98 kPa gauge) but less than or equal to 100 psig (689.5 kPa gauge) shall utilize a testing gauge having increments of 1 psi (6.895 kPa) or less.

3. Tests requiring a pressure greater than 100 psig (689.5 kPa gauge) shall utilize a testing gauge having increments of 2 psi (13.79 kPa) or less.

   Exception: Measuring devices having an equivalent level of accuracy and resolution shall be permitted where specified by the design engineer and approved by the code official.

705.3.7 Test preparation. Pipe joints, including welds, shall be left exposed for examination during the test.

705.3.7.1 Expansion joints. Expansion joints shall be provided with temporary restraints, if required, for the additional thrust load under test.

705.3.7.2 Equipment disconnection. Where the piping system is connected to appliances, equipment or components designed for operating pressures of less than the test pressure, such appliances, equipment and components shall be isolated from the piping system by disconnecting them and capping the outlet(s).

705.3.7.3 Equipment isolation. Where the piping system is connected to appliances, equipment or components designed for operating pressures equal to or greater than the test pressure, such appliances, equipment and components shall be isolated from the piping system by closing the individual appliance, equipment or component shutoff valve(s).

705.4 Detection of leaks and defects. The piping system shall withstand the test pressure specified for the test duration specified without showing any evidence of leakage or other defects. Any reduction of test pressures as indicated by pressure gauges shall indicate a leak within the system. Piping systems shall not be approved except where this reduction in pressure is attributed to some other cause.

705.4.1 Corrections. Where leakage or other defects are identified, the affected portions of the piping system shall be repaired and retested.

705.5 Purging of gaseous hydrogen piping systems. Purging shall comply with Sections 705.5.1 through 705.5.4.

705.5.1 Removal from service. Where piping is to be opened for servicing, addition or modification, the section to be worked on shall be isolated from the supply at the nearest convenient point and the line pressure vented to the outdoors. The remaining gas in this section of pipe shall be displaced with an inert gas.

705.5.2 Placing in operation. Prior to placing the system into operation, the air in the piping system shall be displaced with inert gas. The inert gas flow shall be continued without interruption until the vented gas is free of air. The inert gas shall then be displaced with hydrogen until the vented gas is free of inert gas. The point of discharge shall not be left unattended during purging. After purging, the vent opening shall be closed.

705.5.3 Discharge of purged gases. The open end of piping systems being purged shall not discharge into confined spaces or areas where there are sources of ignition except where precautions are taken to perform this operation in a safe manner by ventilation of the space, control of purging rate and elimination of all hazardous conditions.

705.5.3.1 Vent pipe outlets for purging. Vent pipe outlets for purging shall be located such that the inert gas and fuel gas is released outdoors and not less than 8 feet (2438 mm) above the adjacent ground level. Gases shall be discharged upward or horizontally away from adjacent walls to assist in dispersion. Vent outlets shall be located such that the gas will not be trapped by eaves or other obstructions and shall be at least 5 feet (1524 mm) from building openings and lot lines of properties that can be built upon.

705.5.4 Placing equipment in operation. After the piping has been placed in operation, all equipment shall be purged in accordance with Section 707.2 and then placed in operation, as necessary.

SECTION 706 (IFGC)
LOCATION OF GASEOUS HYDROGEN SYSTEMS

[F] 706.1 General. The location and installation of gaseous hydrogen systems shall be in accordance with Sections 706.2 and 706.3.

Exception: Stationary fuel-cell power plants in accordance with Section 633.

[F] 706.2 Indoor gaseous hydrogen systems. Gaseous hydrogen systems shall be located in indoor rooms or areas in accordance with one of the following:

1. Inside a building in a hydrogen cutoff room designed and constructed in accordance with Section 420 of the *International Building Code*;

2. Inside a building not in a hydrogen cutoff room where the gaseous hydrogen system is listed and labeled for indoor installation and installed in accordance with the manufacturer's installation instructions; and

3. Inside a building in a dedicated hydrogen fuel dispensing area having an aggregate hydrogen delivery capacity not

greater than 12 SCFM and designed and constructed in accordance with Section 703.1.

[F] 706.3 Outdoor gaseous hydrogen systems. Gaseous hydrogen systems shall be located outdoors in accordance with Section 2209.3.2 of the *International Fire Code.*

SECTION 707 (IFGC)
OPERATION AND MAINTENANCE OF
GASEOUS HYDROGEN SYSTEMS

[F] 707.1 Maintenance. Gaseous hydrogen systems and detection devices shall be maintained in accordance with the *International Fire Code* and the manufacturer's installation instructions.

[F] 707.2 Purging. Purging of gaseous hydrogen systems, other than piping systems purged in accordance with Section 705.5, shall be in accordance with Section 2211.8 of the *International Fire Code* or in accordance with the system manufacturer's instructions.

SECTION 708 (IFGC)
DESIGN OF LIQUEFIED HYDROGEN SYSTEMS
ASSOCIATED WITH HYDROGEN
VAPORIZATION OPERATIONS

[F] 708.1 General. The design of liquefied hydrogen systems shall comply with Chapter 32 of the *International Fire Code.*

REFERENCED STANDARDS

This chapter lists the standards that are referenced in various sections of this document. The standards are listed herein by the promulgating agency of the standard, the standard identification, the effective date and title, and the section or sections of this document that reference the standard. The application of the referenced standards shall be as specified in Section 102.8.

ANSI

American National Standards Institute
25 West 43rd Street
Fourth Floor
New York, NY 10036

Standard reference number	Title	Referenced in code section number
ANSI A13.1-96	Scheme for the Identification of piping systems.	704.1.2.2
ANSI CSA-America FC 1-03	Stationery Fuel Cell Power Systems.	633.1
LC 1—97	Interior Gas Piping Systems Using Corrugated Stainless Steel Tubing with Addenda LC1a-1999 and LC1b-2001	403.5.4
Z21.1—03	Household Cooking Gas Appliances with Addenda Z21.1a-2003 and Z21.1b-2003.	623.1
Z21.5.1—02	Gas Clothes Dryers - Volume I -Type 1 Clothes Dryers with Addenda Z21.5.1a-2003.	613.1
Z21.5.2—01	Gas Clothes Dryers - Volume II- Type 2 Clothes Dryers with Addenda Z21.5.2a-2003 and Z21.5.2b-2003.	613.1, 614.3
Z21.8—94 (R2002)	Installation of Domestic Gas Conversion Burners	619.1
Z21.10.1—04	Gas Water Heaters - Volume I - Storage, Water Heaters with Input Ratings of 75,000 Btu per Hour or Less - with Addenda Z21.10.1a-2002	624.1
Z21.10.3—01	Gas Water Heaters - Volume III - Storage, Water Heaters with Input Ratings Above 75,000 Btu per hour, Circulating and Instantaneous – with Addenda Z21.10.3a-2003 and Z21.10.3b-2004	624.1
Z21.11.2—02	Gas-fired Room Heaters - Volume II - Unvented Room Heaters with Addenda Z21.11.2a-2003.	621.1
Z21.13—04	Gas-fired Low-Pressure Steam and Hot Water Boilers.	631.1
Z21.15—97 (R2003)	Manually Operated Gas Valves for Appliances, Appliance Connector Valves, and Hose End Valves with Addenda Z21.15a-2001(R2003).	409.1.1
Z21.19—02	Refrigerators Using Gas (R1999) Fuel	625.1
Z21.24—97	Connectors for Gas Appliances.	411.1
Z21.40.1—96 (R2002)	Gas-fired Heat Activated Air Conditioning and Heat Pump Appliances—with Addendum Z21.40.1a-1997 (R2002)	627.1
Z21.40.2—96 (R2002)	Gas-fired Work Activated Air Conditioning and Heat Pump Appliances (Internal Combustion)—with Addendum Z21.40.2a-97 (R2002)	627.1
Z21.42—93 (R2002)	Gas-fired Illuminating Appliances	628.1
Z21.47—03	Gas-fired Central Furnaces	618.1
Z21.50—03	Vented Gas Fireplaces – with Addenda Z21.50a-2003.	604.1
Z21.56—01	Gas-fired Pool Heaters – with Addenda Z21.56a-2004 and Z21.56b-2004.	616.1
Z21.58—95 (R2002)	Outdoor Cooking Gas Appliances—with Addendum Z21.58a-1998 (R2002) and Z21.58b-2002.	623.1
Z21.60—03	Decorative Gas Appliances for Installation in Solid-fuel Burning Fireplaces – with Addenda Z21.60a-2003	602.1
Z21.61—83 (R 1996)	Toilets, Gas-Fired	625.1
Z21.69—02	Connectors for Movable Gas Appliances – with Addenda Z21.69a-2003.	411.1
Z21.75/CSA 6.27—01	Connectors for Outdoor Gas Appliances and Manufactured Homes.	411.1, 411.2
Z21.80—03	Line Pressure Regulators.	410.1
Z21.84—02	Manually-Lighted, Natural Gas Decorative Gas Appliances for Installation in Solid Fuel Burning Fireplaces – with Addenda Z21.84a-2003	602.1, 602.2
Z21.86—04	Gas-Fired Vented Space Heating Appliances - with Addenda Z21.86a-2002 and Z21.86b-2002	608.1, 609.1, 622.1
Z21.88—02	Vented Gas Fireplace Heaters with Addenda Z21.88a-2003 and Z21.88b-2004.	605.1
Z21.91—01	Ventless Firebox Enclosures for Gas-Fired Unvented Decorative Room Heaters.	621.7.1
Z83.4—03	Non-Recirculating Direct-Gas-Fired Industrial Air Heaters.	611.1
Z83.6—90 (R 1998)	Gas-Fired Infrared Heaters	630.1
Z83.8—02	Gas Unit Heaters and Gas-Fired Duct Furnaces – with Addenda Z83.8a-2003.	620.1
Z83.11—02	Gas Food Service Equipment – with Addenda Z83.11a-2004	623.1
Z83.18—00	Recirculating Direct Gas-Fired Industrial Air Heaters with Addenda Z83.18a 2001 and Z83.18b-2003	612.1

ASME

American Society of Mechanical Engineers
Three Park Avenue
New York, NY 10016-5990

Standard reference number	Title	Referenced in code section number
B1.20.1—83 (Reaffirmed 2001)	Pipe Threads, General Purpose (inch)	403.9
B16.1—98	Cast Iron Pipe Flanges and Flanged Fittings, Class 25, 125 and 250	403.12
B16.20—98	Metallic Gaskets for Pipe Flanges Ring-joint, Spiral-wound, and Jacketed —with Addendum B16.20a-2000	403.12
B31.3-02	Process Piping	704.1.2.4, 704.12, 705.2, 705.3
B16.33—02	Manually Operated Metallic Gas Valves for Use in Gas Piping Systems up to 125 psig (Sizes $^{1}/_{2}$ through 2)	409.1.1
B16.44-01	Manually Operated Metallic Gas Valves for Use in House Piping Systems	409.1.1
B31.3—99	Process Piping	704.1.2, 705.2, 705.3
B36.10M—00	Welded and Seamless Wrought-steel Pipe	403.4.2
BPVC—01	ASME Boiler & Pressure Vessel Code (2001 Edition) (Section I, II, IV, V & IX)	630.1, 703.2.2, 703.3.3, 703.3.4
CSD-1—02	Controls and Safety Devices for Automatically Fired Boilers	631.1

ASTM

ASTM International
100 Barr Harbor Drive
West Conshohocken, PA 19428-2959

Standard reference number	Title	Referenced in code section number
A 53/A 53M—02	Specification for Pipe, Steel, Black and Hot Dipped Zinc-coated Welded and Seamless	403.4.2
A 106—04	Specification for Seamless Carbon Steel Pipe for High-temperature Service	403.4.2
A 254—97 (2002)	Specification for Copper Brazed Steel Tubing	403.5.1
A 539—99	Specification for Electric Resistance-welded Coiled Steel Tubing for Gas and Fuel Oil Lines	403.5.1
B 88—03	Specification for Seamless Copper Water Tube	403.5.2
B 210—02	Specification for Aluminum and Aluminum-alloy Drawn Seamless Tubes	403.5.3
B 241/B 241M—02	Specification for Aluminum and Aluminum-alloy, Seamless Pipe and Seamless Extruded Tube	403.4.4, 403.5.3
B 280—03	Specification for Seamless Copper Tube for Air Conditioning and Refrigeration Field Service	403.5.2
C 64—72 (1977)	Withdrawn No Replacement (Specification for Fireclay Brick Refractories for Heavy Duty Stationary Boiler Service)	503.10.2.5
C 315—02	Specification for Clay Flue Linings	501.12
D 2513—04a	Specification for Thermoplastic Gas Pressure Pipe, Tubing, and Fittings	403.6, 403.6.1, 403.11, 404.14.2

AWWA

American Water Works Association
6666 West Quincy Avenue
Denver, CO 80235

Standard reference number	Title	Referenced in code section number
C111—00	Rubber-Gasket Joints for Ductile-iron Pressure Pipe and Fittings	403.12

CGA

Compressed Gas Association
1725 Jefferson Davis Highway, 5th Floor
Arlington, VA 22202-4102

Standard reference number	Title	Referenced in code section number
S-1.1—(2002)	Pressure Relief Device Standards—Part 1—Cylinders for Compressed Gases	703.3
S-1.2—(1995)	Pressure Relief Device Standards—Part 2—Cargo and Portable Tanks for Compressed Gases	703.3
S-1.3—(1995)	Pressure Relief Device Standards—Part 3—Stationary Storage Containers for Compressed Gases	703.3

CSA

CSA America Inc.
8501 E. Pleasant Valley Rd.
Cleveland, OH USA 44131-5575

Standard reference Part number	Title	Referenced in code section number
ANSI CSA America FC1-03	Stationary Fuel Cell Power Systems	633.1
CSA Requirement 3-88	Manually Operated Gas Valves for Use in House Piping Systems	409.1.1
CSA 8—93	Requirements for Gas-fired Log Lighters for Wood Burning Fireplaces with Revisions through January 1999	603.1

DOTn

Department of Transportation
400 Seventh St. SW.
Washington, DC 20590

Standard reference number	Title	Referenced in code section number
49 CFR,Parts 192.281(e) & 192.283 (b)	Transportation of Natural and Other Gas by Pipeline: Minimum Federal Safety Standards	403.6.1
49 CFR Parts 100-180	Hazardous Materials Regulations	703.2.2, 703.3.3, 703.3.4

ICC

International Code Council, Inc.
500 New Jersey Ave, NW
6th Floor
Washington, DC 20001

Standard reference number	Title	Referenced in code section number
IBC—06	International Building Code®	102.2.1, 201.3, 301.10, 301.11, 301.12, 301.14, 302.1, 302.2, 305.6, 306.6, 401.1.1, 412.6, 413.3, 413.3.1, 501.1, 501.3, 501.12, 501.15.4, 609.3, 614.2, 706.1, 706.3
ICC EC—06	ICC Electrical Code®—Administrative Provisions	201.3, 306.3.1, 306.4.1, 306.5.2, 309.2, 413.8.2.4, 703.6
IEBC—06	International Existing Building Code®	101.2
IECC—06	International Energy Conservation Code®	301.2
IFC—06	International Fire Code®	201.3, 303.4, 401.2, 412.1, 412.6, 412.7, 412.7.3, 412.8, 413.1, 413.3, 413.3.1, 413.4, 413.8.2.5, 701.1, 701.2, 703.2, 703.2.2, 703.3.8, 703.4, 703.5, 704.1.2, 704.3, 704.4, 706.2, 706.3.4, 706.3.5, 706.3.6, 707.2, 707.1, 708.1
IMC—06	International Mechanical Code®	101.2.5, 201.3, 301.1.1, 301.13, 304.11, 501.1, 614.2, 618.5, 621.1, 624.1, 631.2, 632.1, 703.1.2, 706.3.2
IPC—06	International Plumbing Code®	201.3, 301.6, 624.1.1, 624.2
IRC—06	International Residential Code®	703.2.1

MSS

Manufacturers Standardization Society of
the Valve and Fittings Industry
127 Park Street, Northeast
Vienna, VA 22180

Standard reference number	Title	Referenced in code section number
SP-6—01	Standard Finishes for Contact Faces of Pipe Flanges and Connecting-end Flanges of Valves and Fittings	403.12
SP-58—93	Pipe Hangers and Supports—Materials, Design and Manufacture	407.2

NFPA

National Fire Protection Association
1 Batterymarch Pike
P.O. Box 9101
Quincy, MA 02269-9101

Standard reference number	Title	Referenced in code section number
30A-03	Code for Motor Fuel Dispensing Facilities and Repair Garages	305.5
37—02	Installation and Use of Stationary Combustion Engines and Gas Turbines	616.1

NFPA—continued

50A—99	Gaseous Hydrogen Systems at Consumer Sites	706.1
51—02	Design and Installation of Oxygen-Fuel Gas Systems for Welding, Cutting, and Allied Processes	414.1
58—04	Liquefied Petroleum Gas Code	401.2, 402.6.1, 403.6.2, 403.11
82—04	Incinerators, Waste and Linen Handling Systems and Equipment	607.1
85—04	Boiler and Combustion Systems Hazards Code	631.1
211—03	Chimneys, Fireplaces, Vents, and Solid Fuel-burning Appliances	503.5.2, 503.5.3, 503.5.6.1, 503.5.6.3
853—03	Installation of Stationary Fuel Cell Power Systems	633.1

UL

Underwriters Laboratories Inc.
333 Pfingsten Road
Northbrook, IL 60062

Standard reference number	Title	Referenced in code section number
103—2001	Factory-built Chimneys, Residential Type and Building Heating Appliances - with Revisions through December 2003	506.1
127—99	Factory-built Fireplaces—with Revisions through November 1999	621.7
441—96	Gas Vents—with Revisions through December 1999	502.1
641—95	Type L Low-temperature Venting Systems—with Revisions through April 1999	502.1
651—05	Schedule 40 and Schedule 80 Rigid PVC Conduit and Fittings	403.6.3
795—99	Commercial-Industrial Gas Heating Equipment	610.1, 618.1, 631.1
959—01	Medium Heat Appliance Factory-built Chimneys	506.3
1738—00	Venting Systems for Gas Burning Appliances, Categories II, III and IV with Revisions through December 2000	502.1
1777—04	Standard for Chimney Liners	501.12, 501.15.4

APPENDIX A (IFGS)

SIZING AND CAPACITIES OF GAS PIPING

(This appendix is informative and is not part of the code.)

A.1 General piping considerations. The first goal of determining the pipe sizing for a fuel gas piping system is to make sure that there is sufficient gas pressure at the inlet to each appliance. The majority of systems are residential and the appliances will all have the same, or nearly the same, requirement for minimum gas pressure at the appliance inlet. This pressure will be about 5-inch water column (w.c.) (1.25 kPa), which is enough for proper operation of the appliance regulator to deliver about 3.5-inches water column (w.c.) (875 kPa) to the burner itself. The pressure drop in the piping is subtracted from the source delivery pressure to verify that the minimum is available at the appliance.

There are other systems, however, where the required inlet pressure to the different appliances may be quite varied. In such cases, the greatest inlet pressure required must be satisfied, as well as the farthest appliance, which is almost always the critical appliance in small systems.

There is an additional requirement to be observed besides the capacity of the system at 100-percent flow. That requirement is that at minimum flow, the pressure at the inlet to any appliance does not exceed the pressure rating of the appliance regulator. This would seldom be of concern in small systems if the source pressure is $^1/_2$ psi (14-inch w.c.) (3.5 kPa) or less but it should be verified for systems with greater gas pressure at the point of supply.

To determine the size of piping used in a gas piping system, the following factors must be considered:

(1) Allowable loss in pressure from point of delivery to equipment.

(2) Maximum gas demand.

(3) Length of piping and number of fittings.

(4) Specific gravity of the gas.

(5) Diversity factor.

For any gas piping system, or special appliance, or for conditions other than those covered by the tables provided in this code, such as longer runs, greater gas demands or greater pressure drops, the size of each gas piping system should be determined by standard engineering practices acceptable to the code official.

A.2 Description of tables.

A.2.1 General. The quantity of gas to be provided at each outlet should be determined, whenever possible, directly from the manufacturer's gas input Btu/h rating of the appliance that will be installed. In case the ratings of the appliances to be installed are not known, Table 402.2 shows the approximate consumption (in Btu per hour) of certain types of typical household appliances.

To obtain the cubic feet per hour of gas required, divide the total Btu/h input of all appliances by the average Btu heating value per cubic feet of the gas. The average Btu per cubic feet of the gas in the area of the installation can be obtained from the serving gas supplier.

A.2.2 Low pressure natural gas tables. Capacities for gas at low pressure [less than 2.0 psig (13.8 kPa gauge)] in cubic feet per hour of 0.60 specific gravity gas for different sizes and lengths are shown in Tables 402.4(1) and 402.4(2) for iron pipe or equivalent rigid pipe; in Tables 402.4(6) through 402.4(9) for smooth wall semirigid tubing; and in Tables 402.4(13) through 402.4(15) for corrugated stainless steel tubing. Tables 402.4(1) and 402.4(6) are based upon a pressure drop of 0.3-inch w.c. (75 Pa), whereas Tables 402.4(2), 402.4(7) and 402.4(13) are based upon a pressure drop of 0.5-inch w.c. (125 Pa). Tables 402.4(8), 402.4(9), 402.4(14) and 402.4(15) are special low-pressure applications based upon pressure drops greater than 0.5-inch w.c. (125 Pa). In using these tables, an allowance (in equivalent length of pipe) should be considered for any piping run with four or more fittings (see Table A.2.2).

A.2.3 Undiluted liquefied petroleum tables. Capacities in thousands of Btu per hour of undiluted liquefied petroleum gases based on a pressure drop of 0.5-inch w.c. (125 Pa) for different sizes and lengths are shown in Table 402.4(26) for iron pipe or equivalent rigid pipe, in Table 402.4(28) for smooth wall semi-rigid tubing, in Table 402.4(30) for corrugated stainless steel tubing, and in Tables 402.4(33) and 402.4(35) for polyethylene plastic pipe and tubing. Tables 402.4(31) and 402.4(32) for corrugated stainless steel tubing and Table 402.4(34) for polyethylene plastic pipe are based on operating pressures greater than 0.5 pounds per square inch (psi) (3.5 kPa) and pressure drops greater than 0.5-inch w.c. (125 Pa). In using these tables, an allowance (in equivalent length of pipe) should be considered for any piping run with four or more fittings [see Table A.2.2].

A.2.4 Natural gas specific gravity. Gas piping systems that are to be supplied with gas of a specific gravity of 0.70 or less can be sized directly from the tables provided in this code, unless the code official specifies that a gravity factor be applied. Where the specific gravity of the gas is greater than 0.70, the gravity factor should be applied.

Application of the gravity factor converts the figures given in the tables provided in this code to capacities for another gas of different specific gravity. Such application is accomplished by multiplying the capacities given in the tables by the multipliers shown in Table A.2.4. In case the exact specific gravity does not appear in the table, choose the next higher value specific gravity shown.

TABLE A.2.2
EQUIVALENT LENGTHS OF PIPE FITTINGS AND VALVES

		SCREWED FITTINGS[1]				90° WELDING ELBOWS AND SMOOTH BENDS[2]					
		45°/Ell	90°/Ell	180° close return bends	Tee	$R/d = 1$	$R/d = 1\frac{1}{3}$	$R/d = 2$	$R/d = 4$	$R/d = 6$	$R/d = 8$
k factor =		0.42	0.90	2.00	1.80	0.48	0.36	0.27	0.21	0.27	0.36
L/d′ ratio[4] n =		14	30	67	60	16	12	9	7	9	12
Nominal pipe size, inches	Inside diameter d, inches, Schedule 40[6]	L = Equivalent Length In Feet of Schedule 40 (Standard-Weight) Straight Pipe[6]									
$\frac{1}{2}$	0.622	0.73	1.55	3.47	3.10	0.83	0.62	0.47	0.36	0.47	0.62
$\frac{3}{4}$	0.824	0.96	2.06	4.60	4.12	1.10	0.82	0.62	0.48	0.62	0.82
1	1.049	1.22	2.62	5.82	5.24	1.40	1.05	0.79	0.61	0.79	1.05
$1\frac{1}{4}$	1.380	1.61	3.45	7.66	6.90	1.84	1.38	1.03	0.81	1.03	1.38
$1\frac{1}{2}$	1.610	1.88	4.02	8.95	8.04	2.14	1.61	1.21	0.94	1.21	1.61
2	2.067	2.41	5.17	11.5	10.3	2.76	2.07	1.55	1.21	1.55	2.07
$2\frac{1}{2}$	2.469	2.88	6.16	13.7	12.3	3.29	2.47	1.85	1.44	1.85	2.47
3	3.068	3.58	7.67	17.1	15.3	4.09	3.07	2.30	1.79	2.30	3.07
4	4.026	4.70	10.1	22.4	20.2	5.37	4.03	3.02	2.35	3.02	4.03
5	5.047	5.88	12.6	28.0	25.2	6.72	5.05	3.78	2.94	3.78	5.05
6	6.065	7.07	15.2	33.8	30.4	8.09	6.07	4.55	3.54	4.55	6.07
8	7.981	9.31	20.0	44.6	40.0	10.6	7.98	5.98	4.65	5.98	7.98
10	10.02	11.7	25.0	55.7	50.0	13.3	10.0	7.51	5.85	7.51	10.0
12	11.94	13.9	29.8	66.3	59.6	15.9	11.9	8.95	6.96	8.95	11.9
14	13.13	15.3	32.8	73.0	65.6	17.5	13.1	9.85	7.65	9.85	13.1
16	15.00	17.5	37.5	83.5	75.0	20.0	15.0	11.2	8.75	11.2	15.0
18	16.88	19.7	42.1	93.8	84.2	22.5	16.9	12.7	9.85	12.7	16.9
20	18.81	22.0	47.0	105.0	94.0	25.1	18.8	14.1	11.0	14.1	18.8
24	22.63	26.4	56.6	126.0	113.0	30.2	22.6	17.0	13.2	17.0	22.6

continued

TABLE A.2.2—continued
EQUIVALENT LENGTHS OF PIPE FITTINGS AND VALVES

		MITER ELBOWS[3] (No. of miters)					WELDING TEES		VALVES (screwed, flanged, or welded)			
		1-45°	1-60°	1-90°	2-90°[5]	3-90°[5]	Forged	Miter[3]	Gate	Globe	Angle	Swing Check
k factor =		0.45	0.90	1.80	0.60	0.45	1.35	1.80	0.21	10	5.0	2.5
L/d' ratio[4] n =		15	30	60	20	15	45	60	7	333	167	83
Nominal pipe size, inches	Inside diameter d, inches, Schedule 40[6]	L = Equivalent Length In Feet of Schedule 40 (Standard-Weight) Straight Pipe[6]										
1/2	0.622	0.78	1.55	3.10	1.04	0.78	2.33	3.10	0.36	17.3	8.65	4.32
3/4	0.824	1.03	2.06	4.12	1.37	1.03	3.09	4.12	0.48	22.9	11.4	5.72
1	1.049	1.31	2.62	5.24	1.75	1.31	3.93	5.24	0.61	29.1	14.6	7.27
1 1/4	1.380	1.72	3.45	6.90	2.30	1.72	5.17	6.90	0.81	38.3	19.1	9.58
1 1/2	1.610	2.01	4.02	8.04	2.68	2.01	6.04	8.04	0.94	44.7	22.4	11.2
2	2.067	2.58	5.17	10.3	3.45	2.58	7.75	10.3	1.21	57.4	28.7	14.4
2 1/2	2.469	3.08	6.16	12.3	4.11	3.08	9.25	12.3	1.44	68.5	34.3	17.1
3	3.068	3.84	7.67	15.3	5.11	3.84	11.5	15.3	1.79	85.2	42.6	21.3
4	4.026	5.04	10.1	20.2	6.71	5.04	15.1	20.2	2.35	112.0	56.0	28.0
5	5.047	6.30	12.6	25.2	8.40	6.30	18.9	25.2	2.94	140.0	70.0	35.0
6	6.065	7.58	15.2	30.4	10.1	7.58	22.8	30.4	3.54	168.0	84.1	42.1
8	7.981	9.97	20.0	40.0	13.3	9.97	29.9	40.0	4.65	222.0	111.0	55.5
10	10.02	12.5	25.0	50.0	16.7	12.5	37.6	50.0	5.85	278.0	139.0	69.5
12	11.94	14.9	29.8	59.6	19.9	14.9	44.8	59.6	6.96	332.0	166.0	83.0
14	13.13	16.4	32.8	65.6	21.9	16.4	49.2	65.6	7.65	364.0	182.0	91.0
16	15.00	18.8	37.5	75.0	25.0	18.8	56.2	75.0	8.75	417.0	208.0	104.0
18	16.88	21.1	42.1	84.2	28.1	21.1	63.2	84.2	9.85	469.0	234.0	117.0
20	18.81	23.5	47.0	94.0	31.4	23.5	70.6	94.0	11.0	522.0	261.0	131.0
24	22.63	28.3	56.6	113.0	37.8	28.3	85.0	113.0	13.2	629.0	314.0	157.0

For SI: 1 foot = 305 mm, 1 degree = 0.01745 rad.

Note: Values for welded fittings are for conditions where bore is not obstructed by weld spatter or backing rings. If appreciably obstructed, use values for "Screwed Fittings."

1. Flanged fittings have three-fourths the resistance of screwed elbows and tees.
2. Tabular figures give the extra resistance due to curvature alone to which should be added the full length of travel.
3. Small size socket-welding fittings are equivalent to miter elbows and miter tees.
4. Equivalent resistance in number of diameters of straight pipe computed for a value of (f - 0.0075) from the *relation* (n - k/4f).
5. For condition of minimum resistance where the centerline length of each miter is between d and 2 1/2 d.
6. For pipe having other inside diameters, the equivalent resistance may be computed from the above n values.

Source: Crocker, S. *Piping Handbook*, 4th ed., Table XIV, pp. 100-101. Copyright 1945 by McGraw-Hill, Inc. Used by permission of McGraw-Hill Book Company.

TABLE A.2.4
MULTIPLIERS TO BE USED WITH TABLES 402.4(1)
THROUGH 402.4(22) WHERE THE SPECIFIC GRAVITY
OF THE GAS IS OTHER THAN 0.60

SPECIFIC GRAVITY	MULTIPLIER	SPECIFIC GRAVITY	MULTIPLIER
.35	1.31	1.00	.78
.40	1.23	1.10	.74
.45	1.16	1.20	.71
.50	1.10	1.30	.68
.55	1.04	1.40	.66
.60	1.00	1.50	.63
.65	.96	1.60	.61
.70	.93	1.70	.59
.75	.90	1.80	.58
.80	.87	1.90	.56
.85	.84	2.00	.55
.90	.82	2.10	.54

A.2.5 Higher pressure natural gas tables. Capacities for gas at pressures 2.0 psig (13.8 kPa) or greater in cubic feet per hour of 0.60 specific gravity gas for different sizes and lengths are shown in Tables 402.4(3) through 402.4(5) for iron pipe or equivalent rigid pipe; Tables 402.4(10) to 402.4(12) for semirigid tubing; Tables 402.4(16) and 402.4(17) for corrugated stainless steel tubing; and Table 402.4(20) for polyethylene plastic pipe.

A.3 Use of capacity tables.

A.3.1 Longest length method. This sizing method is conservative in its approach by applying the maximum operating conditions in the system as the norm for the system and by setting the length of pipe used to size any given part of the piping system to the maximum value.

To determine the size of each section of gas piping in a system within the range of the capacity tables, proceed as follows (also see sample calculations included in this Appendix):

(1) Divide the piping system into appropriate segments consistent with the presence of tees, branch lines and main runs. For each segment, determine the gas load (assuming all appliances operate simultaneously) and its overall length. An allowance (in equivalent length of pipe) as determined from Table A.2.2 shall be considered for piping segments that include four or more fittings.

(2) Determine the gas demand of each appliance to be attached to the piping system. Where Tables 402.4(1) through 402.4(22) are to be used to select the piping size, calculate the gas demand in terms of cubic feet per hour for each piping system outlet. Where Tables 402.4(23) through 402.4(35) are to be used to select the piping size, calculate the gas demand in terms of thousands of Btu per hour for each piping system outlet.

(3) Where the piping system is for use with other than undiluted liquefied petroleum gases, determine the design

system pressure, the allowable loss in pressure (pressure drop), and specific gravity of the gas to be used in the piping system.

(4) Determine the length of piping from the point of delivery to the most remote outlet in the building/piping system.

(5) In the appropriate capacity table, select the row showing the measured length or the next longer length if the table does not give the exact length. This is the only length used in determining the size of any section of gas piping. If the gravity factor is to be applied, the values in the selected row of the table are multiplied by the appropriate multiplier from Table A.2.4.

(6) Use this horizontal row to locate ALL gas demand figures for this particular system of piping.

(7) Starting at the most remote outlet, find the gas demand for that outlet in the horizontal row just selected. If the exact figure of demand is not shown, choose the next larger figure left in the row.

(8) Opposite this demand figure, in the first row at the top, the correct size of gas piping will be found.

(9) Proceed in a similar manner for each outlet and each section of gas piping. For each section of piping, determine the total gas demand supplied by that section.

When a large number of piping components (such as elbows, tees and valves) are installed in a pipe run, additional pressure loss can be accounted for by the use of equivalent lengths. Pressure loss across any piping component can be equated to the pressure drop through a length of pipe. The equivalent length of a combination of only four elbows/tees can result in a jump to the next larger length row, resulting in a significant reduction in capacity. The equivalent lengths in feet shown in Table A.2.2 have been computed on a basis that the inside diameter corresponds to that of Schedule 40 (standard-weight) steel pipe, which is close enough for most purposes involving other schedules of pipe. Where a more specific solution for equivalent length is desired, this may be made by multiplying the actual inside diameter of the pipe in inches by $n/12$, or the actual inside diameter in feet by n (n can be read from the table heading). The equivalent length values can be used with reasonable accuracy for copper or brass fittings and bends although the resistance per foot of copper or brass pipe is less than that of steel. For copper or brass valves, however, the equivalent length of pipe should be taken as 45 percent longer than the values in the table, which are for steel pipe.

A.3.2 Branch length method. This sizing method reduces the amount of conservatism built into the traditional Longest Length Method. The longest length as measured from the meter to the furthest remote appliance is only used to size the initial parts of the overall piping system. The Branch Length Method is applied in the following manner:

(1) Determine the gas load for each of the connected appliances.

(2) Starting from the meter, divide the piping system into a number of connected segments, and determine the length and amount of gas that each segment would carry assuming that all appliances were operated simul-

taneously. An allowance (in equivalent length of pipe) as determined from Table A.2.2 should be considered for piping segments that include four or more fittings.

(3) Determine the distance from the outlet of the gas meter to the appliance furthest removed from the meter.

(4) Using the longest distance (found in Step 3), size each piping segment from the meter to the most remote appliance outlet.

(5) For each of these piping segments, use the longest length and the calculated gas load for all of the connected appliances for the segment and begin the sizing process in Steps 6 through 8.

(6) Referring to the appropriate sizing table (based on operating conditions and piping material), find the longest length distance in the first column or the next larger distance if the exact distance is not listed. The use of alternative operating pressures and/or pressure drops will require the use of a different sizing table, but will not alter the sizing methodology. In many cases, the use of alternative operating pressures and/or pressure drops will require the approval of both the code official and the local gas serving utility.

(7) Trace across this row until the gas load is found or the closest larger capacity if the exact capacity is not listed.

(8) Read up the table column and select the appropriate pipe size in the top row. Repeat Steps 6, 7 and 8 for each pipe segment in the longest run.

(9) Size each remaining section of branch piping not previously sized by measuring the distance from the gas meter location to the most remote outlet in that branch, using the gas load of attached appliances and following the procedures of Steps 2 through 8.

A.3.3 Hybrid pressure method. The sizing of a 2 psi (13.8 kPa) gas piping system is performed using the traditional Longest Length Method but with modifications. The 2 psi (13.8 kPa) system consists of two independent pressure zones, and each zone is sized separately. The Hybrid Pressure Method is applied as follows:

The sizing of the 2 psi (13.8 kPa) section (from the meter to the line regulator) is as follows:

(1) Calculate the gas load (by adding up the name plate ratings) from all connected appliances. (In certain circumstances the installed gas load may be increased up to 50 percent to accommodate future addition of appliances.) Ensure that the line regulator capacity is adequate for the calculated gas load and that the required pressure drop (across the regulator) for that capacity does not exceed $^3/_4$ psi (5.2 kPa) for a 2 psi (13.8 kPa) system. If the pressure drop across the regulator is too high (for the connected gas load), select a larger regulator.

(2) Measure the distance from the meter to the line regulator located inside the building.

(3) If there are multiple line regulators, measure the distance from the meter to the regulator furthest removed from the meter.

(4) The maximum allowable pressure drop for the 2 psi (13.8 kPa) section is 1 psi (6.9 kPa).

(5) Referring to the appropriate sizing table (based on piping material) for 2 psi (13.8 kPa) systems with a 1 psi (6.9 kPa) pressure drop, find this distance in the first column, or the closest larger distance if the exact distance is not listed.

(6) Trace across this row until the gas load is found or the closest larger capacity if the exact capacity is not listed.

(7) Read up the table column to the top row and select the appropriate pipe size.

(8) If there are multiple regulators in this portion of the piping system, each line segment must be sized for its actual gas load, but using the longest length previously determined above.

The low pressure section (all piping downstream of the line regulator) is sized as follows:

(1) Determine the gas load for each of the connected appliances.

(2) Starting from the line regulator, divide the piping system into a number of connected segments and/or independent parallel piping segments, and determine the amount of gas that each segment would carry assuming that all appliances were operated simultaneously. An allowance (in equivalent length of pipe) as determined from Table A.2.2 should be considered for piping segments that include four or more fittings.

(3) For each piping segment, use the actual length or longest length (if there are sub-branchlines) and the calculated gas load for that segment and begin the sizing process as follows:

(a) Referring to the appropriate sizing table (based on operating pressure and piping material), find the longest length distance in the first column or the closest larger distance if the exact distance is not listed. The use of alternative operating pressures and/or pressure drops will require the use of a different sizing table, but will not alter the sizing methodology. In many cases, the use of alternative operating pressures and/or pressure drops may require the approval of the code official.

(b) Trace across this row until the appliance gas load is found or the closest larger capacity if the exact capacity is not listed.

(c) Read up the table column to the top row and select the appropriate pipe size.

(d) Repeat this process for each segment of the piping system.

A.3.4 Pressure drop per 100 feet method. This sizing method is less conservative than the others, but it allows the designer to immediately see where the largest pressure drop occurs in the system. With this information, modifications can be made to bring the total drop to the critical appliance within the limitations that are presented to the designer.

Follow the procedures described in the Longest Length Method for Steps (1) through (4) and (9).

For each piping segment, calculate the pressure drop based on pipe size, length as a percentage of 100 feet (30 480 mm), and gas flow. Table A.3.4 shows pressure drop per 100 feet (30 480 mm) for pipe sizes from $^1/_2$ inch (12.7 mm) through 2 inch (51 mm). The sum of pressure drops to the critical appliance is subtracted from the supply pressure to verify that sufficient pressure will be available. If not, the layout can be examined to find the high drop section(s) and sizing selections modified.

Note: Other values can be obtained by using the following equation:

$$\text{Desired Value} = MBH \times \sqrt{\frac{\text{Desired Drop}}{\text{Table Drop}}}$$

For example, if it is desired to get flow through $^3/_4$-inch (19.1 mm) pipe at 2 inches/100 feet, multiple the capacity of $^3/_4$-inch pipe at 1 inch/100 feet by the square root of the pressure ratio:

$$147\,MBH \times \sqrt{\frac{2"\,w.c.}{1"\,w.c.}} = 147 \times 1.414 = 208\,MBH$$

$$(MBH = 1000\ Btu/h)$$

A.4 Use of sizing equations. Capacities of smooth wall pipe or tubing can also be determined by using the following formulae:

(1) High Pressure [1.5 psi (10.3 kPa) and above]:

$$Q = 181.6\sqrt{\frac{D^5 \cdot \left(P_1^2 - P_2^2\right) \cdot Y}{C_r \cdot fba \cdot L}}$$

$$= 2237\,D^{2.623}\left[\frac{\left(P_1^2 - P_2^2\right) \cdot Y}{C_r \cdot L}\right]^{0.541}$$

(2) Low Pressure [Less than 1.5 psi (10.3 kPa)]:

$$Q = 187.3\sqrt{\frac{D^5 \cdot \Delta H}{C_r \cdot fba \cdot L}}$$

$$= 2313\,D^{2.623}\left(\frac{\Delta H}{C_r \cdot L}\right)^{0.541}$$

where:

Q = Rate, cubic feet per hour at 60°F and 30-inch mercury column

D = Inside diameter of pipe, in.

P_1 = Upstream pressure, psia

P_2 = Downstream pressure, psia

Y = Superexpansibility factor = 1/supercompressibility factor

C_r = Factor for viscosity, density and temperature*

$$= 0.00354\,ST\left(\frac{Z}{S}\right)^{0.152}$$

Note: See Table 402.4 for Y and C_r for natural gas and propane.

S = Specific gravity of gas at 60°F and 30-inch mercury column (0.60 for natural gas, 1.50 for propane), or = 1488μ

T = Absolute temperature, °F or = $t + 460$

t = Temperature, °F

Z = Viscosity of gas, centipoise (0.012 for natural gas, 0.008 for propane), or = 1488μ

fba = Base friction factor for air at 60°F (CF=1)

L = Length of pipe, ft

ΔH = Pressure drop, in. w.c. (27.7 in. H_2O = 1 psi)

(For SI, see Section 402.4)

A.5 Pipe and tube diameters. Where the internal diameter is determined by the formulas in Section 402.4, Tables A.5.1 and A.5.2 can be used to select the nominal or standard pipe size based on the calculated internal diameter.

TABLE A.3.4
THOUSANDS OF Btu/h (MBH) OF NATURAL GAS PER 100 FEET OF PIPE AT
VARIOUS PRESSURE DROPS AND PIPE DIAMETERS

PRESSURE DROP PER 100 FEET IN INCHES W.C.	PIPE SIZES (inch)					
	$^1/_2$	$^3/_4$	1	$1^1/_4$	$1^1/_2$	2
0.2	31	64	121	248	372	716
0.3	38	79	148	304	455	877
0.5	50	104	195	400	600	1160
1.0	71	147	276	566	848	1640

For SI: 1 inch = 25.4 mm, 1 foot = 304.8 mm.

TABLE A.5.1
SCHEDULE 40 STEEL PIPE STANDARD SIZES

NOMINAL SIZE (in.)	INTERNAL DIAMETER (in.)	NOMINAL SIZE (in.)	INTERNAL DIAMETER (in.)
$^1/_4$	0.364	$1^1/_2$	1.610
$^3/_8$	0.493	2	2.067
$^1/_2$	0.622	$2^1/_2$	2.469
$^3/_4$	0.824	3	3.068
1	1.049	$3^1/_2$	3.548
$1^1/_4$	1.380	4	4.026

For SI: 1 inch = 25.4 mm.

A.6 Use of sizing charts. A third method of sizing gas piping is detailed below as an option that is useful when large quantities of piping are involved in a job (e.g., an apartment house) and material costs are of concern. If the user is not completely familiar with this method, the resulting pipe sizing should be checked by a knowledgeable gas engineer. The sizing charts are applied as follows:

(1) With the layout developed according to Section 106.3.1 of the code, indicate in each section the design gas flow under maximum operation conditions. For many layouts, the maximum design flow will be the sum of all connected loads; However, in some cases, certain combinations of appliances will not occur simultaneously (e.g., gas heating and air conditioning). For these cases, the design flow is the greatest gas flow that can occur at any one time.

(2) Determine the inlet gas pressure for the system being designed. In most cases, the point of inlet will be the gas meter or service regulator, but in the case of a system addition, it could be the point of connection to the existing system.

(3) Determine the minimum pressure required at the inlet to the critical appliance. Usually, the critical item will be the appliance with the highest required pressure for satisfactory operation. If several items have the same required pressure, it will be the one with the greatest length of piping from the system inlet.

(4) The difference between the inlet pressure and critical item pressure is the allowable system pressure drop. Figures A.6(a) and A.6(b) show the relationship between gas flow, pipe size and pipe length for natural gas with 0.60 specific gravity.

(5) To use Figure A.6(a) (low pressure applications), calculate the piping length from the inlet to the critical appliance. Increase this length by 50 percent to allow for fittings. Divide the allowable pressure drop by the equivalent length (in hundreds of feet) to determine the allowable pressure drop per 100 feet (30 480 mm). Select the pipe size from Figure A.6(a) for the required volume of flow.

(6) To use Figure A.6(b) (high pressure applications), calculate the equivalent length as above. Calculate the

index number for Figure A.6(b) by dividing the difference between the squares of the absolute values of inlet and outlet pressures by the equivalent length (in hundreds of feet). Select the pipe size from Figure A.6(b) for the gas volume required.

TABLE A.5.2
COPPER TUBE STANDARD SIZES

TUBE TYPE	NOMINAL OR STANDARD SIZE (inches)	INTERNAL DIAMETER (inches)
K	$^1/_4$	0.305
L	$^1/_4$	0.315
ACR (D)	$^3/_8$	0.315
ACR (A)	$^3/_8$	0.311
K	$^3/_8$	0.402
L	$^3/_8$	0.430
ACR (D)	$^1/_2$	0.430
ACR (A)	$^1/_2$	0.436
K	$^1/_2$	0.527
L	$^1/_2$	0.545
ACR (D)	$^5/_8$	0.545
ACR (A)	$^5/_8$	0.555
K	$^5/_8$	0.652
L	$^5/_8$	0.666
ACR (D)	$^3/_4$	0.666
ACR (A)	$^3/_4$	0.680
K	$^3/_4$	0.745
L	$^3/_4$	0.785
ACR	$^7/_8$	0.785
K	1	0.995
L	1	1.025
ACR	$1^1/_8$	1.025
K	$1^1/_4$	1.245
L	$1^1/_4$	1.265
ACR	$1^3/_8$	1.265
K	$1^1/_2$	1.481
L	$1^1/_2$	1.505
ACR	$1^5/_8$	1.505
K	2	1.959
L	2	1.985
ACR	$2^1/_8$	1.985
K	$2^1/_2$	2.435
L	$2^1/_2$	2.465
ACR	$2^5/_8$	2.465
K	3	2.907
L	3	2.945
ACR	$3^1/_8$	2.945

For SI: 1 inch = 25.4 mm.

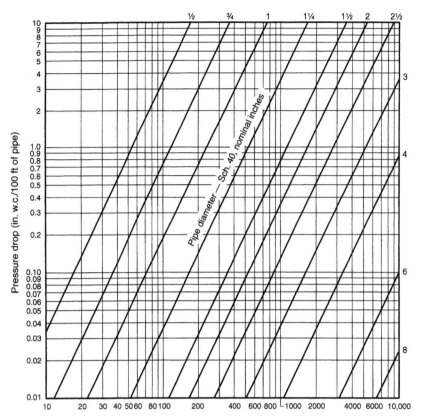

FIGURE A.6 (a)
CAPACITY OF NATURAL GAS PIPING, LOW PRESSURE (0.60 WC)

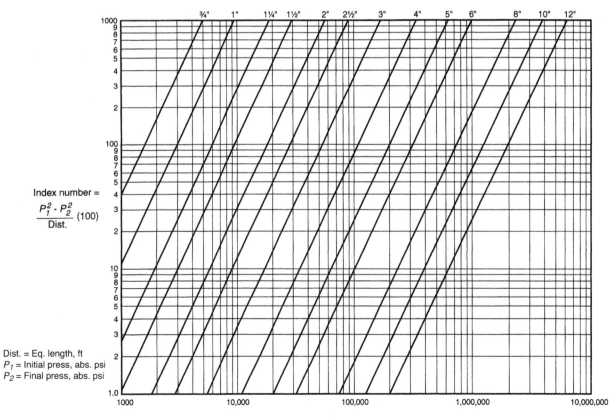

Index number =

$$\frac{P_1^2 - P_2^2}{Dist.}(100)$$

Dist. = Eq. length, ft
P_1 = Initial press, abs. psi
P_2 = Final press, abs. psi

FIGURE A.6 (b)
CAPACITY OF NATURAL GAS PIPING, HIGH PRESSURE (1.5 psi and above)

A.7 Examples of piping system design and sizing

A.7.1 Example 1: Longest length method. Determine the required pipe size of each section and outlet of the piping system shown in Figure A.7.1, with a designated pressure drop of 0.5-inch w.c. (125 Pa) using the Longest Length Method. The gas to be used has 0.60 specific gravity and a heating value of 1,000 Btu/ft³ (37.5 MJ/m³).

Solution:

(1) Maximum gas demand for Outlet A:

$$\frac{\text{Consumption (rating plate input, or Table 402.2 if necessary)}}{\text{Btu of gas}} =$$

$$\frac{35,000 \text{ Btu per hour rating}}{1,000 \text{ Btu per cubic foot}} = 35 \text{ cubic feet per hour} = 35 \text{ cfh}$$

Maximum gas demand for Outlet B:

$$\frac{\text{Consumption}}{\text{Btu of gas}} = \frac{75,000}{1,000} = 75 \text{ cfh}$$

Maximum gas demand for Outlet C:

$$\frac{\text{Consumption}}{\text{Btu of gas}} = \frac{35,000}{1,000} = 35 \text{ cfh}$$

Maximum gas demand for Outlet D:

$$\frac{\text{Consumption}}{\text{Btu of gas}} = \frac{100,000}{1,000} = 100 \text{ cfh}$$

(2) The length of pipe from the point of delivery to the most remote outlet (A) is 60 feet (18 288 mm). This is the only distance used.

(3) Using the row marked 60 feet (18 288 mm) in Table 402.4(2):

    (a) Outlet A, supplying 35 cfh (0.99 m³/hr), requires ¹/₂-inch pipe.

    (b) Outlet B, supplying 75 cfh (2.12 m³/hr), requires ³/₄-inch pipe.

    (c) Section 1, supplying Outlets A and B, or 110 cfh (3.11 m³/hr), requires ³/₄-inch pipe.

    (d) Section 2, supplying Outlets C and D, or 135 cfh (3.82 m³/hr), requires ³/₄-inch pipe.

    (e) Section 3, supplying Outlets A, B, C and D, or 245 cfh (6.94 m³/hr), requires 1-inch pipe.

(4) If a different gravity factor is applied to this example, the values in the row marked 60 feet (18 288 mm) of Table 402.4(2) would be multiplied by the appropriate multiplier from Table A.2.4 and the resulting cubic feet per hour values would be used to size the piping.

A.7.2 Example 2: Hybrid or dual pressure systems. Determine the required CSST size of each section of the piping system shown in Figure A.7.2, with a designated pressure drop of 1 psi (6.9 kPa) for the 2 psi (13.8 kPa) section and 3-inch w.c. (0.75 kPa) pressure drop for the 13-inch w.c. (2.49 kPa) section. The gas to be used has 0.60 specific gravity and a heating value of 1,000 Btu/ft³ (37.5 MJ/ m³).

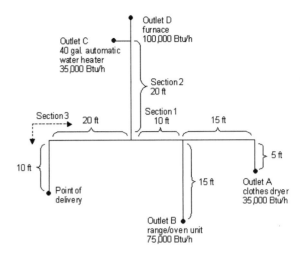

FIGURE A.7.1
PIPING PLAN SHOWING A STEEL PIPING SYSTEM

Solution:

(1) Size 2 psi (13.8 kPa) line using Table 402.4(16).

(2) Size 10-inch w.c. (2.5 kPa) lines using Table 402.4(14).

(3) Using the following, determine if sizing tables can be used.

    (a) Total gas load shown in Figure A.7.2 equals 110 cfh (3.11 m³/hr).

    (b) Determine pressure drop across regulator [see notes in Table 402.4 (16)].

    (c) If pressure drop across regulator exceeds ³/₄ psig (5.2 kPa), Table 402.4 (16) cannot be used. Note: If pressure drop exceeds ³/₄ psi (5.2 kPa), then a larger regulator must be selected or an alternative sizing method must be used.

    (d) Pressure drop across the line regulator [for 110 cfh (3.11 m³/hr)] is 4-inch w.c. (0.99 kPa) based on manufacturer's performance data.

    (e) Assume the CSST manufacturer has tubing sizes or EHDs of 13, 18, 23 and 30.

(4) Section A [2 psi (13.8 kPa) zone]

    (a) Distance from meter to regulator = 100 feet (30 480 mm).

    (b) Total load supplied by A = 110 cfh (3.11 m³/hr) (furnace + water heater + dryer).

    (c) Table 402.4 (16) shows that EHD size 18 should be used.

    Note: It is not unusual to oversize the supply line by 25 to 50 percent of the as-installed load. EHD size 18 has a capacity of 189 cfh (5.35 m³/hr).

(5) Section B (low pressure zone)

    (a) Distance from regulator to furnace is 15 feet (4572 mm).

(b) Load is 60 cfh (1.70 m³/hr).

(c) Table 402.4 (14) shows that EHD size 13 should be used.

(6) Section C (low pressure zone)

(a) Distance from regulator to water heater is 10 feet (3048 mm).

(b) Load is 30 cfh (0.85 m³/hr).

(c) Table 402.4 (14) shows that EHD size 13 should be used.

(7) Section D (low pressure zone)

(a) Distance from regulator to dryer is 25 feet (7620 mm).

(b) Load is 20 cfh (0.57 m³/hr).

(c) Table 402.4(14) shows that EHD size 13 should be used.

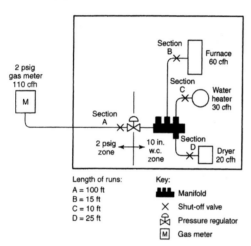

FIGURE A.7.2
PIPING PLAN SHOWING A CSST SYSTEM

A.7.3 Example 3: Branch length method. Determine the required semirigid copper tubing size of each section of the piping system shown in Figure A.7.3, with a designated pressure drop of 1-inch w.c. (250 Pa) (using the Branch Length Method). The gas to be used has 0.60 specific gravity and a heating value of 1,000 Btu/ft³ (37.5 MJ/m³).

Solution:

(1) Section A

(a) The length of tubing from the point of delivery to the most remote appliance is 50 feet (15 240 mm), A + C.

(b) Use this longest length to size Sections A and C.

(c) Using the row marked 50 feet (15 240 mm) in Table 402.4(8), Section A, supplying 220 cfh (6.2 m³/hr) for four appliances requires 1-inch tubing.

(2) Section B

(a) The length of tubing from the point of delivery to the range/oven at the end of Section B is 30 feet (9144 mm), A + B.

(b) Use this branch length to size Section B only.

(c) Using the row marked 30 feet (9144 mm) in Table 402.4(8), Section B, supplying 75 cfh (2.12 m³/hr) for the range/oven requires ¹/₂-inch tubing.

(3) Section C

(a) The length of tubing from the point of delivery to the dryer at the end of Section C is 50 feet (15 240 mm), A + C.

(b) Use this branch length (which is also the longest length) to size Section C.

(c) Using the row marked 50 feet (15 240 mm) in Table 402.4(8), Section C, supplying 30 cfh (0.85 m³/hr) for the dryer requires ³/₈-inch tubing.

(4) Section D

(a) The length of tubing from the point of delivery to the water heater at the end of Section D is 30 feet (9144 mm), A + D.

(b) Use this branch length to size Section D only.

(c) Using the row marked 30 feet (9144 mm) in Table 402.4(8), Section D, supplying 35 cfh (0.99 m³/hr) for the water heater requires ³/₈-inch tubing.

(5) Section E

(a) The length of tubing from the point of delivery to the furnace at the end of Section E is 30 feet (9144 mm), A + E.

(b) Use this branch length to size Section E only.

(c) Using the row marked 30 feet (9144 mm) in Table 402.4(8), Section E, supplying 80 cfh (2.26 m³/hr) for the furnace requires ¹/₂-inch tubing.

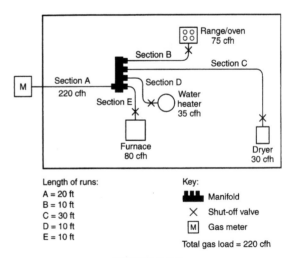

FIGURE A.7.3
PIPING PLAN SHOWING A COPPER TUBING SYSTEM

A.7.4 Example 4: Modification to existing piping system. Determine the required CSST size for Section G (retrofit application) of the piping system shown in Figure A.7.4, with a designated pressure drop of 0.5-inch w.c. (125 Pa) using the branch length method. The gas to be used has 0.60 specific gravity and a heating value of 1,000 Btu/ft³ (37.5 MJ/m³).

Solution:

(1) The length of pipe and CSST from the point of delivery to the retrofit appliance (barbecue) at the end of Section G is 40 feet (12 192 mm), A + B + G.

(2) Use this branch length to size Section G.

(3) Assume the CSST manufacturer has tubing sizes or EHDs of 13, 18, 23 and 30.

(4) Using the row marked 40 feet (12 192 mm) in Table 402.4(13), Section G, supplying 40 cfh (1.13 m³/hr) for the barbecue requires EHD 18 CSST.

(5) The sizing of Sections A, B, F and E must be checked to ensure adequate gas carrying capacity since an appliance has been added to the piping system (see A.7.1 for details).

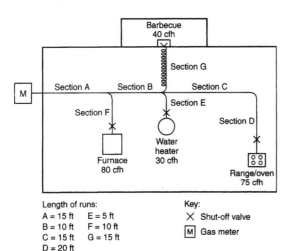

FIGURE A.7.4
PIPING PLAN SHOWING A MODIFICATION
TO EXISTING PIPING SYSTEM

A.7.5 Example 5: Calculating pressure drops due to temperature changes. A test piping system is installed on a warm autumn afternoon when the temperature is 70°F (21°C). In accordance with local custom, the new piping system is subjected to an air pressure test at 20 psig (138 kPa). Overnight, the temperature drops and when the inspector shows up first thing in the morning the temperature is 40°F (4°C).

If the volume of the piping system is unchanged, then the formula based on Boyle's and Charles' law for determining the new pressure at a reduced temperature is as follows:

$$\frac{T_1}{T_2} = \frac{P_1}{P_2}$$

where:

$T_1 =$ Initial temperature, absolute ($T_1 + 459$)

$T_2 =$ Final temperature, absolute ($T_2 + 459$)

$P_1 =$ Initial pressure, psia ($P_1 + 14.7$)

$P_2 =$ Final pressure, psia ($P_2 + 14.7$)

$$\frac{(70 + 459)}{(40 + 459)} = \frac{(20 + 14.7)}{(P_2 + 14.7)}$$

$$\frac{529}{499} = \frac{34.7}{(P_2 + 14.7)}$$

$$(P_2 + 14.7) \times \frac{529}{499} = 34.7$$

$$(P_2 + 14.7) = \frac{34.7}{1.060}$$

$$P_2 = 32.7 - 14.7$$

$$P_2 = 18 \ psig$$

Therefore, the gauge could be expected to register 18 psig (124 kPa) when the ambient temperature is 40°F (4°C).

A7.6 Example 6: Pressure drop per 100 feet of pipe method.
Using the layout shown in Figure A.7.1 and ΔH=pressure drop, in w.c. (27.7 in. H_2O=1 psi), proceed as follows:

(1) Length to A = 20 feet, with 35,000 Btu/hr.

For ¹/₂-inch pipe, $\Delta H = {}^{20 \ feet}/_{100 \ feet} \times 0.3$ inch w.c. = 0.06 in w.c.

(2) Length to B = 15 feet, with 75,000 Btu/hr.

For ³/₄-inch pipe, $\Delta H = {}^{15 \ feet}/_{100 \ feet} \times 0.3$ inch w.c. = 0.045 in w.c.

(3) Section 1 = 10 feet, with 110,000 Btu/hr. Here there is a choice:

For 1 inch pipe: $\Delta H = {}^{10 \ feet}/_{100 \ feet} \times 0.2$ inch w.c. = 0.02 in w.c.

For ³/₄-inch pipe: $\Delta H = {}^{10 \ feet}/_{100 \ feet} \times [0.5$ inch w.c. + ${}^{(110,000 \ Btu/hr-104,000 \ Btu/hr)}/_{(147,000 \ Btu/hr-104,000 \ Btu/hr)} \times (1.0$ inches w.c. - 0.5 inch w.c.)$] = 0.1 \times 0.57$ inch w.c.≈ 0.06 inch w.c.

Note that the pressure drop between 104,000 Btu/hr and 147,000 Btu/hr has been interpolated as 110,000 Btu/hr.

(4) Section 2 = 20 feet, with 135,000 Btu/hr. Here there is a choice:

For 1-inch pipe: $\Delta H = {}^{20 \ feet}/_{100 \ feet} \times [0.2$ inch w.c. + ${}^{(\Delta 14,000 \ Btu/hr)}/_{(\Delta 27,000 \ Btu/hr)} \times \Delta 0.1$ inch w.c.$)] = 0.05$ inch w.c.$)]$

For ³/₄-inch pipe: $\Delta H = {}^{20 \ feet}/_{100 \ feet} \times 1.0$ inch w.c. = 0.2 inch w.c.$)$

Note that the pressure drop between 121,000 Btu/hr and 148,000 Btu/hr has been interpolated as 135,000 Btu/hr, but interpolation for the ¾-inch pipe (trivial for 104,000 Btu/hr to 147,000 Btu/hr) was not used.

(5) Section 3 = 30 feet, with 245,000 Btu/hr. Here there is a choice:

For 1-inch pipe: $\Delta H = {}^{30 \ feet}/_{100 \ feet} \times 1.0$ inches w.c. = 0.3 inch w.c.

For $1^1/_4$-inch pipe: $\Delta H = {}^{30\ feet}/_{100\ feet} \times 0.2$ inch w.c. = 0.06 inch w.c.

Note that interpolation for these options is ignored since the table values are close to the 245,000 Btu/hr carried by that section.

(6) The total pressure drop is the sum of the section approaching A, Sections 1 and 3, or either of the following, depending on whether an absolute minimum is needed or the larger drop can be accommodated.

Minimum pressure drop to farthest appliance:

$\Delta H = 0.06$ inch w.c. $+ 0.02$ inch w.c. $+ 0.06$ inch w.c. $= 0.14$ inch w.c.

Larger pressure drop to the farthest appliance:

$\Delta H = 0.06$ inch w.c. $+ 0.06$ inch w.c. $+ 0.3$ inch w.c. $= 0.42$ inch w.c.

Notice that Section 2 and the run to B do not enter into this calculation, provided that the appliances have similar input pressure requirements.

For SI units: 1 Btu/hr = 0.293 W, 1 cubic foot = 0.028 m³, 1 foot = 0.305 m, 1 inch w.c. = 249 Pa.

SIZING OF VENTING SYSTEMS SERVING APPLIANCES EQUIPPED WITH DRAFT HOODS, CATEGORY I APPLIANCES, AND APPLIANCES LISTED FOR USE WITH TYPE B VENTS

(This appendix is informative and is not part of the code.)

EXAMPLES USING SINGLE APPLIANCE VENTING TABLES

Example 1: Single draft-hood-equipped appliance.

An installer has a 120,000 British thermal unit (Btu) per hour input appliance with a 5-inch-diameter draft hood outlet that needs to be vented into a 10-foot-high Type B vent system. What size vent should be used assuming (a) a 5-foot lateral single-wall metal vent connector is used with two 90-degree elbows, or (b) a 5-foot lateral single-wall metal vent connector is used with three 90-degree elbows in the vent system?

Solution:

Table 504.2(2) should be used to solve this problem, because single-wall metal vent connectors are being used with a Type B vent.

(a) Read down the first column in Table 504.2(2) until the row associated with a 10-foot height and 5-foot lateral is found. Read across this row until a vent capacity greater than 120,000 Btu per hour is located in the

shaded columns labeled "NAT Max" for draft-hood-equipped appliances. In this case, a 5-inch-diameter vent has a capacity of 122,000 Btu per hour and may be used for this application.

(b) If three 90-degree elbows are used in the vent system, then the maximum vent capacity listed in the tables must be reduced by 10 percent (see Section 504.2.3 for single appliance vents). This implies that the 5-inch-diameter vent has an adjusted capacity of only 110,000 Btu per hour. In this case, the vent system must be increased to 6 inches in diameter (see calculations below).

122,000 (.90) = 110,000 for 5-inch vent
From Table 504.2(2), Select 6-inch vent
186,000 (.90) = 167,000; This is greater than the required 120,000. Therefore, use a 6-inch vent and connector where three elbows are used.

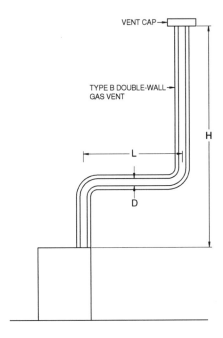

For SI: 1 foot = 304.8 mm, 1 British thermal unit per hour = 0.2931 W.

Table 504.2(1) is used when sizing Type B double-wall gas vent connected directly to the appliance.

Note: The appliance may be either Category I draft hood equipped or fan-assisted type.

FIGURE B-1
TYPE B DOUBLE-WALL VENT SYSTEM SERVING A SINGLE APPLIANCE WITH A TYPE B DOUBLE-WALL VENT

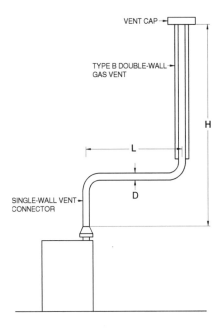

For SI: 1 foot = 304.8 mm, 1 British thermal unit per hour = 0.2931 W.

Table 504.2(2) is used when sizing a single-wall metal vent connector attached to a Type B double-wall gas vent.

Note: The appliance may be either Category I draft hood equipped or fan-assisted type.

FIGURE B-2
TYPE B DOUBLE-WALL VENT SYSTEM SERVING A SINGLE APPLIANCE WITH A SINGLE-WALL METAL VENT CONNECTOR

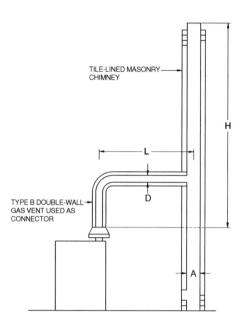

Table 504.2(3) is used when sizing a Type B double-wall gas vent connector attached to a tile-lined masonry chimney.

Note: "A" is the equivalent cross-sectional area of the tile liner.

Note: The appliance may be either Category I draft hood equipped or fan-assisted type.

FIGURE B-3
VENT SYSTEM SERVING A SINGLE APPLIANCE
WITH A MASONRY CHIMNEY OF TYPE B
DOUBLE-WALL VENT CONNECTOR

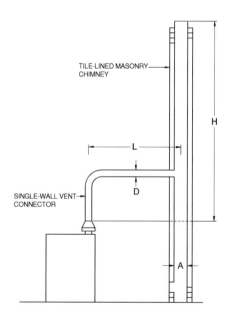

Table 504.2(4) is used when sizing a single-wall vent connector attached to a tile-lined masonry chimney.

Note: "A" is the equivalent cross-sectional area of the tile liner.

Note: The appliance may be either Category I draft hood equipped or fan-assisted type.

FIGURE B-4
VENT SYSTEM SERVING A SINGLE APPLIANCE
USING A MASONRY CHIMNEY AND A
SINGLE-WALL METAL VENT CONNECTOR

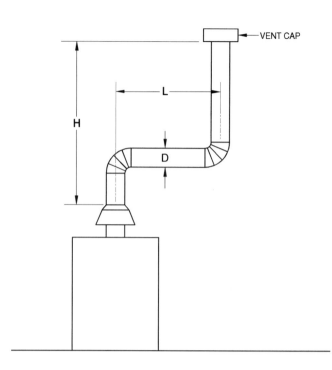

Asbestos cement Type B or single-wall metal vent serving a single draft-hood-equipped appliance [see Table 504.2(5)].

FIGURE B-5
ASBESTOS CEMENT TYPE B OR SINGLE-WALL
METAL VENT SYSTEM SERVING A SINGLE
DRAFT-HOOD-EQUIPPED APPLIANCE

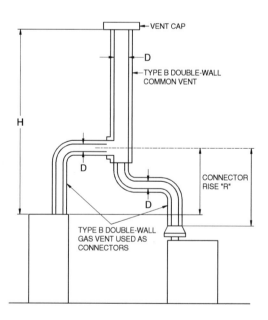

Table 504.3(1) is used when sizing Type B double-wall vent connectors attached to a Type B double-wall common vent.

Note: Each appliance may be either Category I draft hood equipped or fan-assisted type.

FIGURE B-6
VENT SYSTEM SERVING TWO OR MORE APPLIANCES
WITH TYPE B DOUBLE-WALL VENT AND TYPE B
DOUBLE-WALL VENT CONNECTOR

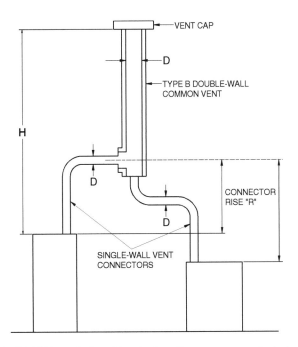

Table 504.3(2) is used when sizing single-wall vent connectors attached to a Type B double-wall common vent.

Note: Each appliance may be either Category I draft hood equipped or fan-assisted type.

FIGURE B-7
VENT SYSTEM SERVING TWO OR MORE APPLIANCES
WITH TYPE B DOUBLE-WALL VENT AND
SINGLE-WALL METAL VENT CONNECTORS

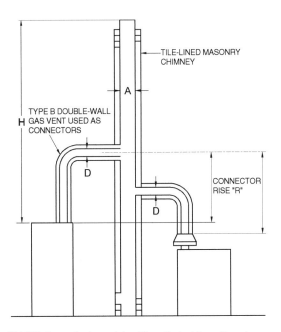

Table 504.3(3) is used when sizing Type B double-wall vent connectors attached to a tile-lined masonry chimney.

Note: "A" is the equivalent cross-sectional area of the tile liner.

Note: Each appliance may be either Category I draft hood equipped or fan-assisted type.

FIGURE B-8
MASONRY CHIMNEY SERVING TWO OR MORE APPLIANCES
WITH TYPE B DOUBLE-WALL VENT CONNECTOR

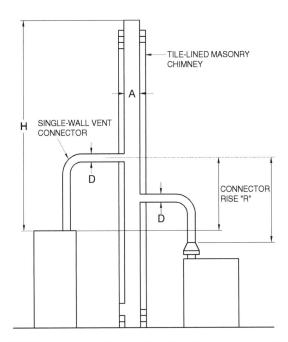

Table 504.3(4) is used when sizing single-wall metal vent connectors attached to a tile-lined masonry chimney.

Note: "A" is the equivalent cross-sectional area of the tile liner.

Note: Each appliance may be either Category I draft hood equipped or fan-assisted type.

FIGURE B-9
MASONRY CHIMNEY SERVING TWO OR MORE APPLIANCES
WITH SINGLE-WALL METAL VENT CONNECTORS

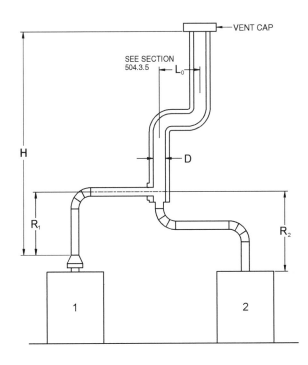

Asbestos cement Type B or single-wall metal pipe vent serving two or more draft-hood-equipped appliances [see Table 504.3(5)].

FIGURE B-10
ASBESTOS CEMENT TYPE B OR SINGLE-WALL
METAL VENT SYSTEM SERVING TWO OR MORE
DRAFT-HOOD-EQUIPPED APPLIANCES

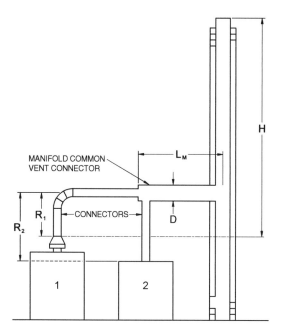

Example: Manifolded Common Vent Connector L_M shall be no greater than 18 times the common vent connector manifold inside diameter; i.e., a 4-inch (102 mm) inside diameter common vent connector manifold shall not exceed 72 inches (1829 mm) in length (see Section 504.3.4).

Note: This is an illustration of a typical manifolded vent connector. Different appliance, vent connector, or common vent types are possible. Consult Section 502.3.

FIGURE B-11
USE OF MANIFOLD COMMON VENT CONNECTOR

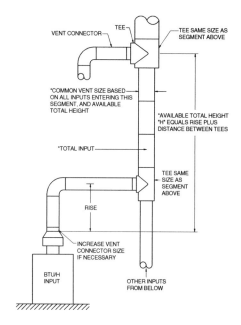

Vent connector size depends on:
- Input
- Rise
- Available total height "H"
- Table 504.3(1) connectors

Common vent size depends on:
- Combined inputs
- Available total height "H"
- Table 504.3(1) common vent

FIGURE B-13
MULTISTORY GAS VENT DESIGN PROCEDURE FOR EACH SEGMENT OF SYSTEM

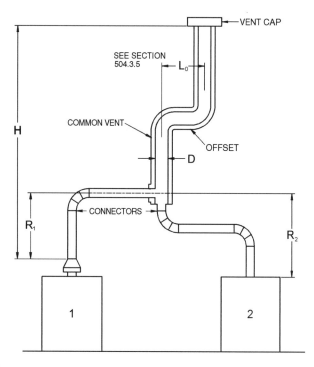

Example: Offset Common Vent

Note: This is an illustration of a typical offset vent. Different appliance, vent connector, or vent types are possible. Consult Sections 504.2 and 504.3.

FIGURE B-12
USE OF OFFSET COMMON VENT

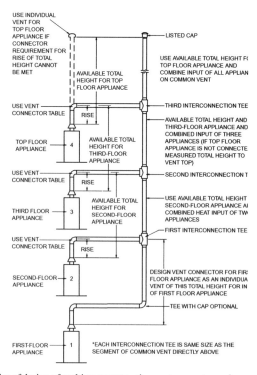

Principles of design of multistory vents using vent connector and common vent design tables (see Sections 504.3.11 through 504.3.17).

FIGURE B-14
MULTISTORY VENT SYSTEMS

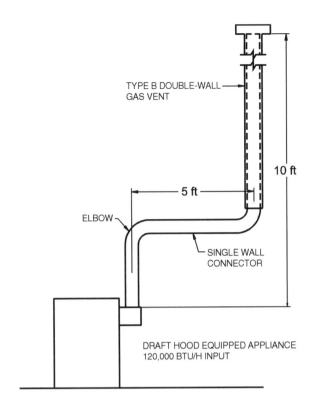

For SI: 1 foot = 304.8 mm, 1 British thermal unit per hour = 0.2931 W.

FIGURE B-15 (EXAMPLE 1)
SINGLE DRAFT-HOOD-EQUIPPED APPLIANCE

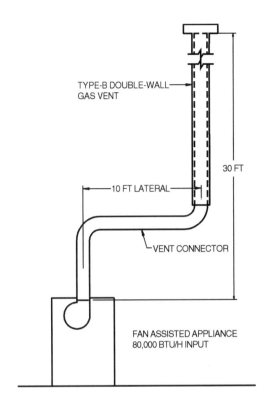

For SI: 1 foot = 304.8 mm, 1 British thermal unit per hour = 0.2931 W.

FIGURE B-16 (EXAMPLE 2)
SINGLE FAN-ASSISTED APPLIANCE

Example 2: Single fan-assisted appliance.

An installer has an 80,000 Btu per hour input fan-assisted appliance that must be installed using 10 feet of lateral connector attached to a 30-foot-high Type B vent. Two 90-degree elbows are needed for the installation. Can a single-wall metal vent connector be used for this application?

Solution:

Table 504.2(2) refers to the use of single-wall metal vent connectors with Type B vent. In the first column find the row associated with a 30-foot height and a 10-foot lateral. Read across this row, looking at the FAN Min and FAN Max columns, to find that a 3-inch-diameter single-wall metal vent connector is not recommended. Moving to the next larger size single wall connector (4 inches), note that a 4-inch-diameter single-wall metal connector has a recommended minimum vent capacity of 91,000 Btu per hour and a recommended maximum vent capacity of 144,000 Btu per hour. The 80,000 Btu per hour fan-assisted appliance is outside this range, so the conclusion is that a single-wall metal vent connector cannot be used to vent this appliance using 10 feet of lateral for the connector.

However, if the 80,000 Btu per hour input appliance could be moved to within 5 feet of the vertical vent, then a 4-inch single-wall metal connector could be used to vent the appliance. Table 504.2(2) shows the acceptable range of vent capacities for a 4-inch vent with 5 feet of lateral to be between 72,000 Btu per hour and 157,000 Btu per hour.

If the appliance cannot be moved closer to the vertical vent, then Type B vent could be used as the connector material. In this case, Table 504.2(1) shows that for a 30-foot-high vent with 10 feet of lateral, the acceptable range of vent capacities for a 4-inch-diameter vent attached to a fan-assisted appliance is between 37,000 Btu per hour and 150,000 Btu per hour.

Example 3: Interpolating between table values.

An installer has an 80,000 Btu per hour input appliance with a 4-inch-diameter draft hood outlet that needs to be vented into a 12-foot-high Type B vent. The vent connector has a 5-foot lateral length and is also Type B. Can this appliance be vented using a 4-inch-diameter vent?

Solution:

Table 504.2(1) is used in the case of an all Type B vent system. However, since there is no entry in Table 504.2(1) for a height of 12 feet, interpolation must be used. Read down the 4-inch diameter NAT Max column to the row associated with 10-foot height and 5-foot lateral to find the capacity value of 77,000 Btu per hour. Read further down to the 15-foot height, 5-foot lateral row to find the capacity value of 87,000 Btu per hour. The difference between the 15-foot height capacity value and the 10-foot height capacity value is 10,000 Btu per hour. The capacity for a vent system with a 12-foot height is equal to the capacity for a 10-foot height plus $^2/_5$ of the difference between the 10-foot and 15-foot height values, or $77,000 + ^2/_5 (10,000) = 81,000$ Btu per hour. Therefore, a 4-inch-diameter vent may be used in the installation.

EXAMPLES USING COMMON VENTING TABLES

Example 4: Common venting two draft-hood-equipped appliances.

A 35,000 Btu per hour water heater is to be common vented with a 150,000 Btu per hour furnace using a common vent with a total height of 30 feet. The connector rise is 2 feet for the water heater with a horizontal length of 4 feet. The connector rise for the furnace is 3 feet with a horizontal length of 8 feet. Assume single-wall metal connectors will be used with Type B vent. What size connectors and combined vent should be used in this installation?

Solution:

Table 504.3(2) should be used to size single-wall metal vent connectors attached to Type B vertical vents. In the vent connector capacity portion of Table 504.3(2), find the row associated with a 30-foot vent height. For a 2-foot rise on the vent connector for the water heater, read the shaded columns for draft-hood-equipped appliances to find that a 3-inch-diameter vent connector has a capacity of 37,000 Btu per hour. Therefore, a 3-inch single-wall metal vent connector may be used with the water heater. For a draft-hood-equipped furnace with a 3-foot rise, read across the appropriate row to find that a 5-inch-diameter vent connector has a maximum capacity of 120,000 Btu per hour (which is too small for the furnace) and a 6-inch-diameter vent connector has a maximum vent capacity of 172,000 Btu per hour. Therefore, a 6-inch-diameter vent connector should be used with the 150,000 Btu per hour furnace. Since both vent connector horizontal lengths are less than the maximum lengths listed in Section 504.3.2, the table values may be used without adjustments.

In the common vent capacity portion of Table 504.3(2), find the row associated with a 30-foot vent height and read over to the NAT + NAT portion of the 6-inch-diameter column to find a maximum combined capacity of 257,000 Btu per hour. Since the two appliances total only 185,000 Btu per hour, a 6-inch common vent may be used.

Example 5a: Common venting a draft-hood-equipped water heater with a fan-assisted furnace into a Type B vent.

In this case, a 35,000 Btu per hour input draft-hood-equipped water heater with a 4-inch-diameter draft hood outlet, 2 feet of connector rise, and 4 feet of horizontal length is to be common vented with a 100,000 Btu per hour fan-assisted furnace with a 4-inch-diameter flue collar, 3 feet of connector rise, and 6 feet of horizontal length. The common vent consists of a 30-foot height of Type B vent. What are the recommended vent diameters for each connector and the common vent? The installer would like to use a single-wall metal vent connector.

Solution: - [Table 504.3(2)]

Water Heater Vent Connector Diameter. Since the water heater vent connector horizontal length of 4 feet is less than the maximum value listed in Section 504.3.2, the venting table values may be used without adjustments. Using the Vent Connector Capacity portion of Table 504.3(2), read down the Total Vent Height (H) column to 30 feet and read across the 2-foot Connector Rise (R) row to the first Btu per hour rating in the NAT Max column that is equal to or greater than the water heater input rating. The table shows that a 3-inch vent connector has a maximum input rating of 37,000 Btu per hour. Although this is greater than the water heater input rating, a 3-inch vent connector is prohibited by Section 504.3.21. A 4-inch vent connector

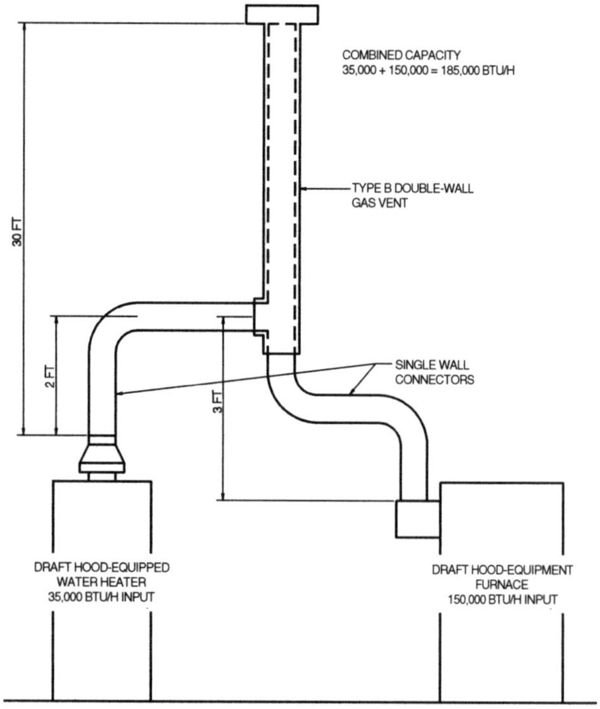

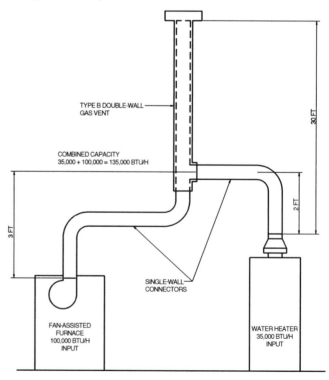

FIGURE B-17 (EXAMPLE 4)
COMMON VENTING TWO DRAFT-
HOOD-EQUIPPED APPLIANCES

FIGURE B-18 (EXAMPLE 5A)
COMMON VENTING A DRAFT HOOD WITH A FAN-ASSISTED
FURNACE INTO A TYPE B DOUBLE-WALL COMMON VENT

has a maximum input rating of 67,000 Btu per hour and is equal to the draft hood outlet diameter. A 4-inch vent connector is selected. Since the water heater is equipped with a draft hood, there are no minimum input rating restrictions.

Furnace Vent Connector Diameter. Using the Vent Connector Capacity portion of Table 504.3(2), read down the Total Vent Height (*H*) column to 30 feet and across the 3-foot Connector Rise (*R*) row. Since the furnace has a fan-assisted combustion system, find the first FAN Max column with a Btu per hour rating greater than the furnace input rating. The 4-inch vent connector has a maximum input rating of 119,000 Btu per hour and a minimum input rating of 85,000 Btu per hour. The 100,000 Btu per hour furnace in this example falls within this range, so a 4-inch connector is adequate. Since the furnace vent connector horizontal length of 6 feet does not exceed the maximum value listed in Section 504.3.2, the venting table values may be used without adjustment. If the furnace had an input rating of 80,000 Btu per hour, then a Type B vent connector [see Table 504.3(1)] would be needed in order to meet the minimum capacity limit.

Common Vent Diameter. The total input to the common vent is 135,000 Btu per hour. Using the Common Vent Capacity portion of Table 504.3(2), read down the Total Vent Height (*H*) column to 30 feet and across this row to find the smallest vent diameter in the FAN + NAT column that has a Btu per hour rating equal to or greater than 135,000 Btu per hour. The 4-inch common vent has a capacity of 132,000 Btu per hour and the 5-inch common vent has a capacity of 202,000 Btu per hour. Therefore, the 5-inch common vent should be used in this example.

Summary. In this example, the installer may use a 4-inch-diameter, single-wall metal vent connector for the water heater and a 4-inch-diameter, single-wall metal vent connector for the furnace. The common vent should be a 5-inch-diameter Type B vent.

Example 5b: Common venting into a masonry chimney.

In this case, the water heater and fan-assisted furnace of Example 5a are to be common vented into a clay tile-lined masonry chimney with a 30-foot height. The chimney is not exposed to the outdoors below the roof line. The internal dimensions of the clay tile liner are nominally 8 inches by 12 inches. Assuming the same vent connector heights, laterals, and materials found in Example 5a, what are the recommended vent connector diameters, and is this an acceptable installation?

Solution:

Table 504.3(4) is used to size common venting installations involving single-wall connectors into masonry chimneys.

Water Heater Vent Connector Diameter. Using Table 504.3(4), Vent Connector Capacity, read down the Total Vent Height (*H*) column to 30 feet, and read across the 2-foot Connector Rise (*R*) row to the first Btu per hour rating in the NAT Max column that is equal to or greater than the water heater input rating. The table shows that a 3-inch vent connector has a maximum input of only 31,000 Btu per hour while a 4-inch vent connector has a maximum input of 57,000 Btu per hour. A 4-inch vent connector must therefore be used.

Furnace Vent Connector Diameter. Using the Vent Connector Capacity portion of Table 504.3(4), read down the Total Vent Height (*H*) column to 30 feet and across the 3-foot Connector Rise (*R*) row. Since the furnace has a fan-assisted combustion system, find the first FAN Max column with a Btu per hour rating greater than the furnace input rating. The 4-inch vent connector has a maximum input rating of 127,000 Btu per hour and a minimum input rating of 95,000 Btu per hour. The 100,000 Btu per hour furnace in this example falls within this range, so a 4-inch connector is adequate.

Masonry Chimney. From Table B-1, the equivalent area for a nominal liner size of 8 inches by 12 inches is 63.6 square inches. Using Table 504.3(4), Common Vent Capacity, read down the FAN + NAT column under the Minimum Internal Area of Chimney value of 63 to the row for 30-foot height to find a capacity value of 739,000 Btu per hour. The combined input rating of the furnace and water heater, 135,000 Btu per hour, is less than the table value, so this is an acceptable installation.

Section 504.3.17 requires the common vent area to be no greater than seven times the smallest listed appliance categorized vent area, flue collar area, or draft hood outlet area. Both appliances in this installation have 4-inch-diameter outlets. From Table B-1, the equivalent area for an inside diameter of 4 inches is 12.2 square inches. Seven times 12.2 equals 85.4, which is greater than 63.6, so this configuration is acceptable.

Example 5c: Common venting into an exterior masonry chimney.

In this case, the water heater and fan-assisted furnace of Examples 5a and 5b are to be common vented into an exterior masonry chimney. The chimney height, clay tile liner dimensions, and vent connector heights and laterals are the same as in Example 5b. This system is being installed in Charlotte, North Carolina. Does this exterior masonry chimney need to be relined? If so, what corrugated metallic liner size is recommended? What vent connector diameters are recommended?

Solution:

According to Section 504.3.20, Type B vent connectors are required to be used with exterior masonry chimneys. Use Table 504.3(7) to size FAN+NAT common venting installations involving Type-B double wall connectors into exterior masonry chimneys.

The local 99-percent winter design temperature needed to use Table 504.3(7) can be found in the ASHRAE *Handbook of Fundamentals*. For Charlotte, North Carolina, this design temperature is 19°F.

Chimney Liner Requirement. As in Example 5b, use the 63 square inch Internal Area columns for this size clay tile liner. Read down the 63 square inch column of Table 504.3(7a) to the 30-foot height row to find that the combined appliance maximum input is 747,000 Btu per hour. The combined input rating of the appliances in this installation, 135,000 Btu per hour, is less than the maximum value, so this criterion is satisfied. Table 504.3(7b), at a 19°F design temperature, and at the same vent height and internal area used above, shows that the minimum allowable input rating of a space-heating appliance is 470,000 Btu per hour. The furnace input rating of 100,000 Btu per hour is less than this minimum value. So this criterion is not satisfied, and an alternative venting design needs to be used, such as a Type B vent shown in Example 5a or a listed chimney liner system shown in the remainder of the example.

According to Section 504.3.19, Table 504.3(1) or 504.3(2) is used for sizing corrugated metallic liners in masonry chimneys, with the maximum common vent capacities reduced by 20 percent. This example will be continued assuming Type B vent connectors.

Water Heater Vent Connector Diameter. Using Table 504.3(1), Vent Connector Capacity, read down the Total Vent Height (H) column to 30 feet, and read across the 2-foot Connector Rise (R) row to the first Btu/h rating in the NAT Max column that is equal to or greater than the water heater input rating. The table shows that a 3-inch vent connector has a maximum capacity of 39,000 Btu/h. Although this rating is greater than the water heater input rating, a 3-inch vent connector is prohibited by Section 504.3.21. A 4-inch vent connector has a maximum input rating of 70,000 Btu/h and is equal to the draft hood outlet diameter. A 4-inch vent connector is selected.

Furnace Vent Connector Diameter. Using Table 504.3(1), Vent Connector Capacity, read down the Vent Height (H) column to 30 feet, and read across the 3-foot Connector Rise (R) row to the first Btu per hour rating in the FAN Max column that is equal to or greater than the furnace input rating. The 100,000 Btu per hour furnace in this example falls within this range, so a 4-inch connector is adequate.

Chimney Liner Diameter. The total input to the common vent is 135,000 Btu per hour. Using the Common Vent Capacity Portion of Table 504.3(1), read down the Vent Height (H) column to 30 feet and across this row to find the smallest vent diameter in the FAN+NAT column that has a Btu per hour rating greater than 135,000 Btu per hour. The 4-inch common vent has a capacity of 138,000 Btu per hour. Reducing the maximum capacity by 20 percent (Section 504.3.19) results in a maximum capacity for a 4-inch corrugated liner of 110,000 Btu per hour, less than the total input of 135,000 Btu per hour. So a larger liner is needed. The 5-inch common vent capacity listed in Table 504.3(1) is 210,000 Btu per hour, and after reducing by 20 percent is 168,000 Btu per hour. Therefore, a 5-inch corrugated metal liner should be used in this example.

Single-Wall Connectors. Once it has been established that relining the chimney is necessary, Type B double-wall vent connectors are not specifically required. This example could be redone using Table 504.3(2) for single-wall vent connectors. For this case, the vent connector and liner diameters would be the same as found above with Type B double-wall connectors.

TABLE B-1
MASONRY CHIMNEY LINER DIMENSIONS
WITH CIRCULAR EQUIVALENTS[a]

NOMINAL LINER SIZE (inches)	INSIDE DIMENSIONS OF LINER (inches)	INSIDE DIAMETER OR EQUIVALENT DIAMETER (inches)	EQUIVALENT AREA (square inches)
4 × 8	$2^1/_2 \times 6^1/_2$	4	12.2
		5	19.6
		6	28.3
		7	38.3
8 × 8	$6^3/_4 \times 6^3/_4$	7.4	42.7
		8	50.3
8 × 12	$6^1/_2 \times 10^1/_2$	9	63.6
		10	78.5
12 × 12	$9^3/_4 \times 9^3/_4$	10.4	83.3
		11	95
12 × 16	$9^1/_2 \times 13^1/_2$	11.8	107.5
		12	113.0
		14	153.9
16 × 16	$13^1/_4 \times 13^1/_4$	14.5	162.9
		15	176.7
16 × 20	13 × 17	16.2	206.1
		18	254.4
20 × 20	$16^3/_4 \times 16^3/_4$	18.2	260.2
		20	314.1
20 × 24	$16^1/_2 \times 20^1/_2$	20.1	314.2
		22	380.1
24 × 24	$20^1/_4 \times 20^1/_4$	22.1	380.1
		24	452.3
24 × 28	$20^1/_4 \times 20^1/_4$	24.1	456.2
28 × 28	$24^1/_4 \times 24^1/_4$	26.4	543.3
		27	572.5
30 × 30	$25^1/_2 \times 25^1/_2$	27.9	607
		30	706.8
30 × 36	$25^1/_2 \times 31^1/_2$	30.9	749.9
		33	855.3
36 × 36	$31^1/_2 \times 31^1/_2$	34.4	929.4
		36	1017.9

For SI: 1 inch = 25.4 mm, 1 square inch = 645.16 m².

a. Where liner sizes differ dimensionally from those shown in Table B-1, equivalent diameters may be determined from published tables for square and rectangular ducts of equivalent carrying capacity or by other engineering methods.

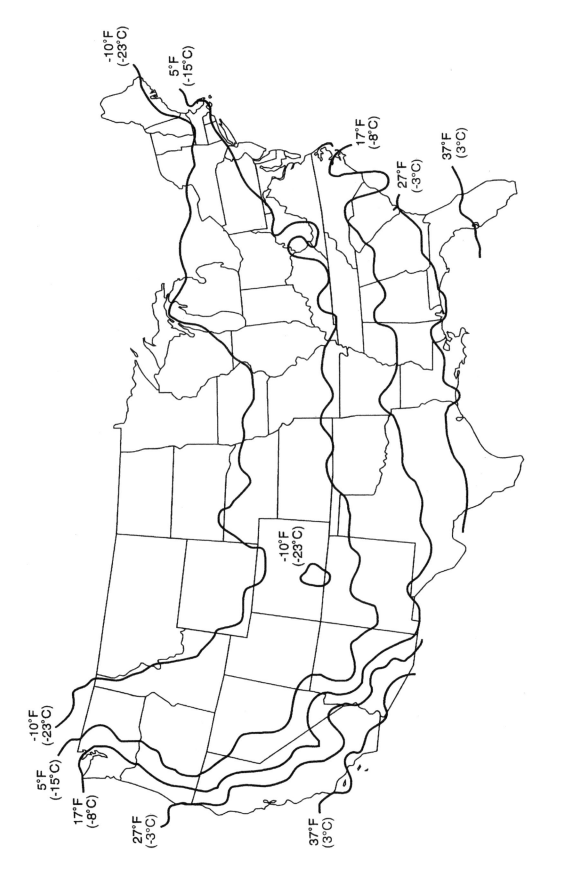

-10°F
(-23°C)

5°F
(-15°C)

17°F
(-8°C)

27°F
(-3°C)

37°F
(3°C)

-10°F
(-23°C)

5°F   -10°F
(-15°C) (-23°C)

17°F
(-8°C)

27°F
(-3°C)

37°F
(3°C)

FIGURE B-19

APPENDIX C (IFGS)

EXIT TERMINALS OF MECHANICAL DRAFT AND DIRECT-VENT VENTING SYSTEMS

(This appendix is informative and is not part of the code.)

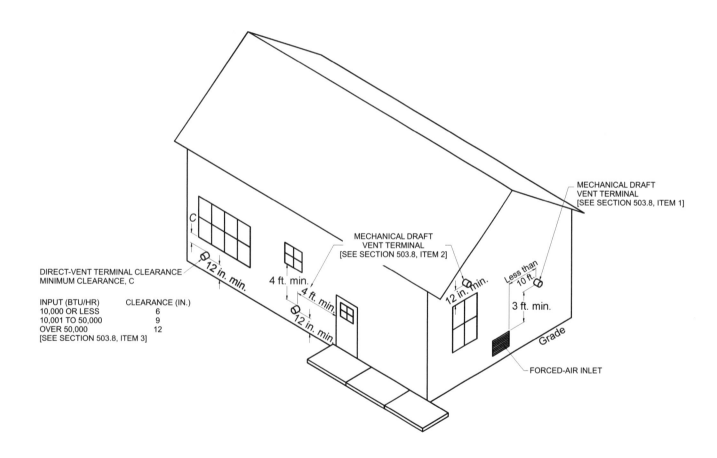

For SI: 1 inch = 25.4 mm, 1 foot = 304.8 mm, 1 British thermal unit per hour = 0.2931 W.

APPENDIX C
EXIT TERMINALS OF MECHANICAL DRAFT AND DIRECT-VENT VENTING SYSTEMS

APPENDIX D (IFGS)

RECOMMENDED PROCEDURE FOR SAFETY INSPECTION OF AN EXISTING APPLIANCE INSTALLATION

(This appendix is informative and is not part of the code.)

The following procedure is intended as a guide to aid in determining that an appliance is properly installed and is in a safe condition for continuing use.

This procedure is predicated on central furnace and boiler installations, and it should be recognized that generalized procedures cannot anticipate all situations. Accordingly, in some cases, deviation from this procedure is necessary to determine safe operation of the equipment.

(a) This procedure should be performed prior to any attempt at modification of the appliance or of the installation.

(b) If it is determined there is a condition that could result in unsafe operation, the appliance should be shut off and the owner advised of the unsafe condition. The following steps should be followed in making the safety inspection:

1. Conduct a check for gas leakage. (See Section 406.6)

2. Visually inspect the venting system for proper size and horizontal pitch and determine there is no blockage or restriction, leakage, corrosion, and other deficiencies that could cause an unsafe condition.

3. Shut off all gas to the appliance and shut off any other fuel-gas-burning appliance within the same room. **Use the shutoff valve in the supply line to each appliance.**

4. Inspect burners and crossovers for blockage and corrosion.

5. **Applicable only to furnaces.** Inspect the heat exchanger for cracks, openings, or excessive corrosion.

6. **Applicable only to boilers.** Inspect for evidence of water or combustion product leaks.

7. Insofar as is practical, close all building doors and windows and all doors between the space in which the appliance is located and other spaces of the building. Turn on clothes dryers. Turn on any exhaust fans, such as range hoods and bathroom exhausts, so they will operate at maximum speed. Do not operate a summer exhaust fan. Close fireplace dampers. If, after completing Steps 8 through 13, it is believed sufficient combustion air is not available, refer to Section 304 of this code for guidance.

8. Place the appliance being inspected in operation. **Follow the lighting instructions.** Adjust the thermostat so appliance will operate continuously.

9. Determine that the pilot(s), where provided, is burning properly and that the main burner ignition is satisfactory by interrupting and reestablishing the electrical supply to the appliance in any convenient manner. If the appliance is equipped with a continuous pilot(s), test the pilot safety device(s) to determine if it is operating properly by extinguishing the pilot(s) when the main burner(s) is off and determining, after 3 minutes, that the main burner gas does not flow upon a call for heat. If the appliance is not provided with a pilot(s), test for proper operation of the ignition system in accordance with the appliance manufacturer's lighting and operating instructions.

10. Visually determine that the main burner gas is burning properly (i.e., no floating, lifting, or flashback). Adjust the primary air shutter(s) as required. If the appliance is equipped with high and low flame controlling or flame modulation, check for proper main burner operation at low flame.

11. Test for spillage at the draft hood relief opening after 5 minutes of main burner operation. Use a flame of a match or candle or smoke.

12. Turn on all other fuel-gas-burning appliances within the same room so they will operate at their full inputs. **Follow lighting instructions for each appliance.**

13. Repeat Steps 10 and 11 on the appliance being inspected.

14. Return doors, windows, exhaust fans, fireplace dampers, and any other fuel-gas-burning appliance to their previous conditions of use.

15. **Applicable only to furnaces.** Check both the limit control and the fan control for proper operation. Limit control operation can be checked by blocking the circulating air inlet or temporarily disconnecting the electrical supply to the blower motor and determining that the limit control acts to shut off the main burner gas.

16. **Applicable only to boilers.** Determine that the water pumps are in operating condition. Test low water cutoffs, automatic feed controls, pressure and temperature limit controls, and relief valves in accordance with the manufacturer's recommendations to determine that they are in operating condition.

INDEX

W

EDITORIAL CHANGES-SECOND PRINTING

Page 123, Table 634.1: line 4, column 8 now reads . . . 80,000

Page 138, Table A.2.4: table title, third line now reads . . . THROUGH 402.4(22) WHERE THE SPECIFIC GRAVITY

Page 134, UL, new standard added now reads . . . 651—05 Schedule 40 and Schedule 80 Rigid PVC Conduit and Fittings 403.6.3

INTERNATIONAL CODE COUNCIL®

Membership Application
This form may be photocopied

MEMBERSHIP CATEGORIES AND DUES*— ANNUAL MEMBERSHIP

Special membership structures are also available for Educational Institutions and Federal Agencies.
For more information, please visit www.iccsafe.org/membership or call 1-888-ICC-SAFE (422-7233), x33804.

GOVERNMENTAL MEMBER**
Government/Municipality (including agencies, departments or units) engaged in administration, formulation or enforcement of laws, regulations or ordinances relating to public health, safety and welfare. Annual member dues (by population) are shown below. Please verify the current ICC membership status of your employer prior to applying.

☐ Up to 50,000..........$100   ☐ 50,001–150,000........ $180   ☐ 150,001+.......... $280

**A Governmental Member may designate 4 to 12 voting representatives (based on population) who are employees or officials of that governmental member and are actively engaged on a full- or part-time basis in the administration, formulation or enforcement of laws, regulations or ordinances relating to public health, safety and welfare. Number of representatives is based on population. All dues for representatives have been included in the annual member dues payment. Please call 1-888-ICC-SAFE (422-7233), x33804 for information about how to designate your voting representatives.

☐ **CORPORATE MEMBERS ($300)**   An association, society, testing laboratory, manufacturer, company or corporation

INDIVIDUAL MEMBERS

☐ **PROFESSIONAL ($150)**   A design professional duly licensed or registered by any state or other recognized governmental agency

☐ **COOPERATING ($150)**   An individual who is interested in International Code Council purposes and objectives and would like to take advantage of membership benefits

☐ **CERTIFIED ($75)**   An individual who holds a current Legacy or International Code Council certification

☐ **ASSOCIATE ($35)**   An employee of a current ICC Governmental Member

☐ **STUDENT ($25)**   An individual who is enrolled in classes or a course of study including at least 12 hours of classroom instruction per week

☐ **RETIRED ($20)**   A former governmental representative, corporate or individual member who has retired

☐ **TEMPORARY ($50)**   A trial 4-month non-renewing membership with all of the benefits provided to our individuals members, except a complimentary code book.

New Governmental and Corporate Members will receive a free package of 7 code books. New Individual Members will receive one free code book. Upon receipt of your completed application and payment, you will be contacted by an ICC Member Services Representative regarding your free code package or code book. For more information, please visit www.iccsafe.org/membership or call 1-888-ICC-SAFE (422-7233), x33804.

Please print clearly or type information below:

_____
Name

_____
Name of Jurisdiction, Association, Institute or Company, etc.

_____
Title

_____
Billing Address

_____
City   State   Zip+4

_____
Street Address for Shipping

_____
City   State   Zip+4

_____
E-mail

_____
Telephone

_____
Tax Exempt Number (If applicable, must attach copy of tax exempt license if claiming an exemption)
Payment Information:

_____
VISA, MC, AMEX or DISCOVER Account Number   **Exp. Date**

Return this application to:
International Code Council
Attn: Membership
5360 Workman Mill Road
Whittier, CA 90601-2298

Toll Free: 1-888-ICC-SAFE (1-888-422-7233), x33804
FAX: (562) 692-6031 (Los Angeles District Office)
Or, apply online at **www.iccsafe.org/membership.**
Please refer to Tracking Number 66-07-027 when applying.

If you have any questions about membership in the International Code Council, call 1-888-ICC-SAFE (1-888-422-7233), x33804 and request a Member Services Representative.

REF 66-07-027

*Membership categories and dues subject to change.
Please visit www.iccsafe.org/membership for the most current information.